Welcome to the Machine

Welcome to the Machine

Science, Surveillance, and the Culture of Control

Derrick Jensen
George Draffan

A POLITICS OF THE LIVING BOOK

CHELSEA GREEN PUBLISHING
WHITE RIVER JUNCTION, VERMONT

Managing Editor: Collette Leonard
Project Editor: Marcy Brant
Developmental Editor: Helen Whybrow
Copy Editor: Lori Lewis
Proofreader: Susannah Noel
Indexer: Peggy Holloway
Designer: Peter Holm

Printed in Canada

First printing, July 2004

10 9 8 7 6 5 4 3 2 1

Printed on acid-free, recycled paper

Library of Congress Cataloging-in-Publication Data

Jensen, Derrick, 1960-
  Welcome to the machine : science, surveillance, and the culture of control
/ Derrick Jensen and George Draffan.
      p. cm. – (Politics of the living)
  Includes bibliographical references and index.
  ISBN 1-931498-52-0 (pbk.)
  1. Computers and civilization–Congresses. 2. Privacy, Right
of–Congresses. I. Draffan, George, 1954- II. Title. III. Series.
  QA76.9.C66J48 2004
  303.48'34–dc22
                                                    2004007833

Chelsea Green Publishing Company
Post Office Box 428
White River Junction, VT 05001
(800) 639-4099
www.chelseagreen.com

Contents

The All-Seeing Eye • 1
Science • 14
Control • 33
Identity • 50
The Machine • 57
Fear • 78
Rationalization • 88
The Panoptic Sort • 112
Nothing to Fear • 125
The Real World • 142
Money • 147
The Noose Tightens • 168
The End • 187
Humanity • 205

Acknowledgments • 227
Notes • 229
Bibliography • 257
Index • 277
About the Authors • 286

Because this book is in part about identity, it is important that we be clear about ours. The narrator of this book is Derrick, so the pronoun *I* refers to him. *We* refers to Derrick and George, or to collective humans.

In the past the man has been first; in the future the System must be first.

Frederick Winslow Taylor,
founder of scientific management, 1911

I should like merely to understand how it happens that so many men, so many villages, so many cities, so many nations, sometimes suffer under a single tyrant who has no other power than the power they give him; who is able to harm them only to the extent to which they have the willingness to bear with him; who could do them absolutely no injury unless they preferred to put up with him rather than contradict him.

Éttiene de La Boétie, 1564

The All-Seeing Eye

The eye you see isn't an eye because you see it; it's an
eye because it sees you.

Antonio Machado

When I was a child, I was taught—as a fundamentalist
Christian—that while the devil could not read my mind, he
watched everything I did, scanning for the slightest shift of my
body or expression that would reveal my thoughts. He did this, I
was told, because he wanted to know me. And he wanted to
know me not because he loved me—as God did, who watched me
also and who knew in addition what went on in my head and in
my heart—but because he wanted to tempt and even control me.

My response as a child was to attempt to control myself, to let
neither my face nor body, nor especially my actions, reveal my
thoughts. I'd fool him! But I knew even at age five that this was
a waste of time. I knew—though of course I could not have used
this language—that if the devil, or for that matter anyone, could
assemble a large enough body of data about my external habits, he
could in time effectively read my mind. I knew also that the
capacity to read my mind, whether by God, man, or devil, would
lead necessarily to the capacity to control me: surveillance controls, and absolute surveillance controls absolutely.

What I didn't realize at the time was that by attempting to
control myself I was effectively surrendering my freedom. I was
allowing my fear—of the devil, and in retrospect even more so of
God—to determine my actions, my expressions, my thoughts,
and most damning of all, what I did not think.

I no longer believe in a devil, nor in a God, at least not the sort
about which I was taught as a child. I do, however, carry with me

the lessons I learned about the relationship between information held by a distant authority and control by that authority. This relationship has always been understood by those in power. It is a relationship we all need to remember.

By now, most of us can see the central movement of our culture: for the last several thousand years it has relentlessly expanded its region of control from its original base in ancient Mesopotamia—the "cradle of civilization"—through the Middle East and Levant, around the Mediterranean, into Europe, then Africa, the Americas, Asia, Oceania. In exerting this control, the culture has deforested more than 90 percent of the world, depleted more than 90 percent of the world's fisheries, similarly destroyed the great flocks of birds, the great herds of ungulates. It has destroyed, subsumed, or forcibly assimilated nearly all the cultures in its path, until most of these other ways of perceiving and being in the world have been forgotten. This much is clear. These are simply facts. They are beyond dispute.[1]

It would be a mistake, however, to think that this movement toward the attempt at absolute control extends only into the external world. It extends as surely into our inner worlds, into what we think and who we are, with the attempted control as complete, the devastation as severe, as that in the outer world.

Pretend you're sitting in a room, surrounded by cops. Or perhaps you're surrounded by representatives of a major corporation. Sometimes you can't tell the difference.

They ask you questions, show you pictures, read you slogans. You do not want to respond. You do not trust them, these cops, these representatives of a corporation. You do not want to give them information.

But you do. Not by speaking. Not by a barely perceptible tightening of your clasped hands, nor by a shifting of your shoulders.

You do not betray yourself by a flash of recognition, nor even by the slightest movement of your eyes nor a moistening of your skin with sweat. Instead you are betrayed by the activities inside your brain itself. They have a machine that can read these activities. And there's nothing you can do about it. They know this. They get the information they want.

Or you walk into a store. The representative of a major corporation—or maybe it's a cop, sometimes you can't tell the difference—glances at a console, then greets you by name. You do not know this representative, this cop. You have never seen him before in your life. But he knows you. He knows where you bought your shirt, and how much it cost. He knows the same about your pants, socks, shoes, backpack, car. He may know your credit history, your medical conditions, your arrest (or non-arrest) record.

It's not you. He's not reading your brain, as did the cops—or maybe they were representatives of a major corporation. He's reading your clothes. He's reading every mass-produced item in your possession. He's reading your cash. The items contain computer chips, you see. Or rather, they hold chips you *don't* see. The chips are smaller than a grain of sand. And cheap, five cents a piece, soon down to a penny. These chips continuously broadcast information about you to whoever has the proper receiver. Agents of major corporations—or of the state, whichever they happen to be—know how much you paid for your sweater, know where you are, and know where you've been.

You begin to notice cameras everywhere you go. At first they were only in obvious places, like casinos and 7-11s. But then you start seeing them elsewhere, in intersections, ATMs, hotel hallways, schools. You notice them at airports, and even at the Super Bowl. No matter where you are, someone is able to watch you, record your movements. Sometimes these watchers—cops, or maybe salesmen from major corporations—use computers to scan your face and compare these scans to other faces in their databases, maybe deadbeats, maybe criminals, maybe terrorists, maybe people whose politics they don't like, maybe people who do or don't buy their products.

Or maybe you're crazy. Paranoid. You've been hearing voices lately. You can't always tell which thoughts are yours, and which belong to someone else. You're not even sure any longer who you are. Who are you? Do you think your thoughts, or are they someone else's?

You've read that scientists at Yale have been doing studies in which they use magnetic resonance imaging machines (MRIs) that provide pictures of what physically goes on inside the brain, to detect people's responses to pictures. (They found, no big surprise here, that many white people who claim not to be racist, and who in fact may not be particularly racist on a conscious level, feel fear when shown pictures of unfamiliar black men.) And you know MRIs are already used by market research companies to "gauge consumers' unconscious preferences by looking at the pattern of brain activity as they respond to products or messages."[2] And if you *know* that market researchers are using MRIs, then you probably can't even imagine the uses the cops, Federal Bureau of Investigation (FBI), and Central Intelligence Agency (CIA) have dreamed up. One clue: researchers at the University of Pennsylvania have discovered that some parts of your brain "light up" (to use their words) when you are presented with a face you've seen before.[3] Another clue: scientists at Stanford are beginning to map the electrical and magnetic brain waves associated with *thinking* specific words and sentences. You think *this,* your brain does *that*. Already, the scientists say, "Recognition rates, based on a least-squares criterion, varied, but the best were above 90 percent."[4]

You have the right to remain silent. It just won't do you any good.

You know, too, about the computer chips. They're everywhere. Not yet. But soon. Katherine Albrecht wrote recently in the *Denver University Law Review,* "A new consumer goods tracking system called Auto-ID [now called RFID, for Radio Frequency Identification tags] is poised to enter all of our lives, with profound implications for consumer privacy. Auto-ID couples radio frequency (RF) identification technology with highly miniaturized computers that enable products to be identified and tracked at any point along the supply chain. The system could be applied to almost any physical item, from ballpoint pens to toothpaste, which would carry their own unique information in the form of an embedded chip. The chip sends out an identification signal allowing it to communicate with reader devices and other products embedded with similar chips. Analysts envision a time when the system will be used to identify and track every item produced on the planet. Auto-ID employs a numbering scheme called ePC (for 'electronic product code') which can provide a unique ID for any physical object in the world. . . . For example, each pack of cigarettes, individual can of soda, light bulb or package of razor blades produced would be uniquely identifiable through its own ePC number. Once assigned, this number is transmitted by a radio frequency ID tag (RFID) in or on the product. These tiny tags, predicted by some to cost less than 1 cent each by 2004, are 'somewhere between the size of a grain of sand and a speck of dust.' They are to be built directly into food, clothes, drugs, or auto-parts during the manufacturing process. Receiver or reader devices are used to pick up the signal transmitted by the RFID tag. Proponents envision a pervasive global network of millions of receivers along the entire supply chain—in airports, seaports, highways, distribution centers, warehouses, retail stores, and in the home. This would allow for seamless, continuous identification and tracking of physical items as they move from one place to another, enabling companies to determine the whereabouts of all their products at all times. Steven Van Fleet, an executive at International Paper, looks forward to the prospect. 'We'll put a radio frequency ID tag on everything that moves in the North American supply chain,' he enthused recently. The ultimate goal is for Auto-ID to create a 'physically linked world' in which every item on the planet is numbered, identified, catalogued, and tracked. And the technology exists to make this a reality. Described as 'a political rather than a technological problem,' creating a global system 'would . . . involve negotiation between, and consensus among, different countries.' Supporters are aiming for worldwide acceptance of the technologies needed to build the infrastructure within the next few years."[5]

The Auto-ID Center at the Massachusetts Institute of Technology (MIT) has funding and/or participation from global corporations and universities in the United States, United Kingdom, Australia, Japan, and Switzerland.[6] Sponsors include Pepsi, Gillette, Philip Morris, Procter & Gamble, Wal-Mart, and (who would have guessed?) the U.S. Department of Defense.[7] In 2001 this gang wired the city of Tulsa, Oklahoma with RFID to see if they could track objects tagged with Auto-ID.[8] Gillette, Wal-Mart, and Tesco are installing shelves that can read RFID radio waves embedded in shavers and related products.[9] The European Central Bank is planning to imbed Euro banknotes with RFID tags by 2005. Hitachi Europe has already developed a smart tag chip small enough, at .3 mm square and thin as a human hair, to fit inside a banknote. Mass production of these chips will start within a year.[10]

And the cameras. You've seen them with your own eyes. You know. You know.

But maybe you don't. There are still the voices, telling you that you're crazy, saying you're paranoid, saying it's no big deal, saying those in power have more important things to do than to watch you (more important things like tracking the movements of every roll of toilet paper, every dollar bill, every person: NO! Don't think like that! That's not right!). The voices tell you that those in power have your best interests at heart. The chips are to reduce theft, the cameras to increase security, and the MRIs, well, if you have nothing to hide, what are you afraid of?

Another article, written not so long ago, began with the unforgettable first line: "Those voices in your head may be real." It went on to say that scientists have been able to develop the capacity to project a beam of sound so focused that only one person can hear it. It can be transmitted from hundreds of yards away. The military is of course extremely interested in this technology.[11] Microwaves can also be used to transmit sound. Pulses can be beamed into your head, such that you might think that you're hearing them, or even thinking them. These pulses could be shaped into words, into thoughts.[12]

But that's all crazy, the voices tell you. None of that could happen. None of that could happen to you.

Just today I learned that schoolchildren in Akron, Ohio are being fingerprinted so they can be identified when they go through school lunch lines.[13] This seems pretty absurd to me. Lunchroom monitors "back in the day" had little trouble keeping track of who was little Derrick and who was little George, with no need for fingerprints or photo IDs. The monitors knew who we were and often had relationships with us: imagine that! And even if groups of children from other schools (or, the horror, home-schooled children!) had snuck into lunchrooms so they could steal taxpayer-subsidized mystery meat, I can't see how it would have been cost effective (or relationship effective) for the government to go all high-tech on their ass.

I suspect, however, that there are other reasons for the fingerprinting. Especially if the students are children of color, it may be useful for the police to keep the prints on file for later use: it saves trouble when figuring out how to send them to prison. But I'm sure there's another reason, too, that works for all children in our culture, regardless of race: forcing children to endure day after day of sitting in rows bored out of their skulls breaks their wills and destroys their intelligence sufficiently to prepare them to take their proper places in adult society, where they will lead lives of drudgery, obedience, tedium, painful employment, and quiet desperation. As R. D. Laing wrote, "Children are not yet fools, but we shall turn them into imbeciles like ourselves, with high IQ's if possible."[14] So too this forced fingerprinting prepares children for adulthood, in this case for lives in which they will submit to more or less constant surveillance.

Say hello to the twenty-first century.

One of the pioneers of modern surveillance was the eighteenth-century utilitarian philosopher Jeremy Bentham, designer of the Panopticon. The Panopticon is a blueprint for a prison designed as a cylinder, with cells radiating from the central guard station. There are no nooks or crannies where prisoners can hide. The cells are always lit, while the guard station is dark. Because prisoners can never tell whether or when they are being watched, they have no choice but to presume that at every moment they are under surveillance.

Here is what, with the Panopticon, Bentham proposed to accomplish: "Morals reformed—health preserved—industry invigorated instruction diffused—public burthens lightened—Economy seated, as it were, upon a rock—the gordian knot of the Poor-Laws are not cut, but untied—all by a simple idea in Architecture!"[15] Perhaps more to the point, whoever ran the Panopticon would gain a "new mode of obtaining power of mind over mind, in a quantity hitherto without example."[16]

Bentham was ambitious. This power was to be used widely,

for "*punishing the incorrigible, guarding the insane, reforming the vicious, confining the suspected, employing the idle, maintaining the helpless, curing the sick, instructing the willing* in any branch of industry, or *training the rising race* in the path of *education*: in a word, whether it be applied to the purposes of *perpetual prisons* in the room of death, or *prisons for confinement* before trial, or *penitentiary-houses, or houses of correction, or work-houses, or manufactories, or mad-houses, or hospitals, or schools.*"[17]

Here's how it works: "It is obvious that, in all these instances, the more constantly the persons to be inspected are under the eyes of the persons who should inspect them, the more perfectly will the purpose X of the establishment have been attained. Ideal perfection, if that were the object, would require that each person should actually be in that predicament, during every instant of time. This being impossible, the next thing to be wished for is, that, at every instant, seeing reason to believe as much, and not being able to satisfy himself to the contrary, he should *conceive* himself to be so."[18]

The Panopticon serves as the model for modern supermaximum security prisons such as Pelican Bay in Crescent City, California. But Bentham's ideas have been even more influential than that.

Indeed, as Michel Foucault wrote in the 1970s, the Panopticon has become a model for the entire culture. The Panopticon has become not only a "simple idea in Architecture," but also a metaphor for the power relations that undergird modern civilization: "Hence," Foucault wrote, "the major effect of the Panopticon: to induce in the inmate a state of conscious and permanent visibility that assures the automatic functioning of power. So to arrange things that the surveillance is permanent in its effects, even if it is discontinuous in its action; that the perfection of power should tend to render its actual exercise unnecessary; that this architectural apparatus should be a machine for creating and sustaining a power relation independent of the person who exercises it; in short, that the inmates should be caught up in a power situation of which they are themselves the bearers. To achieve this, it is at once too much and too little that the prisoner should be constantly observed by an inspector: too little, for what matters is that he knows himself to be observed; too much, because he has no need in fact of being so. In view of this, Bentham laid down the principle that power should be visible and unverifiable. Visible: the inmate will constantly have before his eyes the tall outline of the central tower from which he is spied upon. Unverifiable: the inmate must never know whether he is being looked at at any one moment; but he must be sure that he may always be so. In order to make the presence or absence of the inspector unverifiable, so that the prisoners, in their cells, cannot even see a shadow, Bentham envisaged not only venetian blinds on the windows of the central observation hall, but, on the inside, partitions that intersected the hall at right angles and, in order to pass from one quarter to the other, not doors but zig-zag openings; for the slightest noise, a gleam of light, a brightness in a half-opened door would betray the presence of the guardian. The Panopticon is a machine for dissociating the see/being seen dyad: in the peripheric ring, one is totally seen, without ever seeing; in the central tower, one sees everything without ever being seen."[19]

That is bad enough, but Foucault continues, "It is an important mechanism, for it automizes and disindividualizes power. Power has its principle not so much in a person as in a certain concerted distribution of bodies, surfaces, lights, gazes; in an arrangement whose internal mechanisms produce the relation in which individuals are caught up. . . . There is a machinery that assures dissymmetry, disequilibrium, difference. Consequently, it does not matter who exercises power."[20]

When I was very young, I sometimes heard the phrase, "money is the root of all evil." When old enough to look it up myself, I found that the quote was inaccurate. Paul (or at least the authors of the King James Bible) had actually said that the *love of* money was the problem, and I grew quite fond of correcting my elders on this score. Now as an adult I've come full circle and see that my childhood understanding was closer to the point. The only word I quibble with is *all,* as I think we need to leave room for the evil caused by patriarchy, industrialism, nation-states, military technologies, vivisection, Christianity and other institutional religion (including the apostle Paul himself), numbness, and the designated hitter rule.

Money. I remember how when I was a kid the backs of dollar bills used to creep me out. To the left of where it says "In God We Trust" (how weird, I thought, to have that phrase on something that is the root of all evil?), there was the obverse of the Great Seal of the United States, with its truncated pyramid and its all-seeing eye. Combine my understanding that money is the root of all evil with my belief that the devil could see my every move, and you might get a sense of why I never put an upside-down dollar bill on my nightstand before I went to sleep.

I've since learned that the eye on the back of the dollar bill does not belong to the devil, but, essentially as scary, to God, or in the secular language of the U.S. State Department, to providence. It is there because, as the Latin phrase *ANNUIT CŒPTIS* written over the eye in this Great Seal means, "the eye of providence has favored our undertakings." George Washington, whose face, of course, appears on the front of the bill, put it well (if a tad awkwardly) in his inaugural address: "No people can be bound to acknowledge and adore the Invisible Hand, which conducts the Affairs of men more than the People of the United States. Every step, by which they have advanced to the character of an independent nation, seems to have been distinguished by some token of providential agency."[21]

Certainly the Revolutionary War came at a providential

moment for Washington, one of the richest men in the new United States. His fortune—and of course we can say the same for essentially all fortunes, but this was especially true in his case—was founded on land illegally taken from Indians. In fact, his fortune was so fraudulent that the (British) governor of Virginia declared Washington's title to these lands null and void. "Providentially" the Revolutionary War broke out the same month as this declaration, saving Washington's fortune.[22]

I'm not sure whether the Indians, whose lands were being stolen by Washington and many others, would be so quick to ascribe the loss of their lands and their way of life—what is these days called genocide—so much to the Invisible Hand of providence as it was to an entire culture of rapacious people hellbent on taking everything that could be turned into money and destroying everything they did not understand.

Annuit Cœptus. "The eye of providence has favored our under-takings."

You and I both know who determines what actions are deemed "providential," that is, ordained by the all-seeing eye of God, and what actions are deemed improvidential, that is, subject to the all-seeing eye of the guard at the center of the Panopticon.

Similarly, when those in power put on the back of the dollar bill "In God We Trust," we can have a pretty good idea who this *we* is, and what we can trust this God—as mediated by those in power—to do.

The controller of the Panopticon thus becomes god, or rather God. Omniscient God. Invisible God. God with oversight both omnipossible and unverifiable. God unapproachable and unknowable. God mediated and represented by those in power. God so internalized that we would never even conceive of going against His word. Or is it the word of those in power? Power is God. Control is the way to God. Those in control are God.

Perhaps the fear of God was once enough to keep people in

line, but for better or worse we live now in what we perceive to be a more secular society. And what happens if those of us in our cells begin to disbelieve that there is someone at the center of the Panopticon, that we are no longer watched? What will we do? How will we act? How will we act if we no longer believe in original sin, no longer believe we deserve to be in these cells, under the watchful eye of those in power? What if we no longer believe we need their supervision? What will we do then?

And what will those in power do? How will they shift, and whom will they bring in to replace the Old Man who may no longer be there?

Or perhaps that's the wrong way to look at it. Perhaps the Old God at the center of the Panopticon never left at all but merely changed form. Perhaps we are seen now as clearly as we ever have been. Perhaps the cells are even smaller, the lights brighter, the space outside our cells ever darker.

Science

Science and technology constitute two major oppressions of our time. Yet, if one goes by the literature, not only are science and technology seen as liberators (either from superstition, fear or material deprivation and want), those who control and direct them (technocrats, industrialists, statists) are seen as liberators too.

Claude Alvares

Of course our culture today is not secular, but just as religious (in the pejorative sense of superstitious, unconscious, assumed) as ever. Only today, science is the religion, experts are the priests, bureaucrats are the gatekeepers, and research and development institutions are the cathedrals.

Right now, military researchers at MIT and elsewhere are working hard to fabricate technologies that will—and we have to stress that we're not making this up—allow soldiers to leap buildings, deflect bullets, and even become invisible. Shoes containing power packs will store energy when soldiers—or state police, or corporate security guards, insofar as there's a difference—walk, then release this energy in bursts to allow them to jump over walls. Soldiers—cops, corporate goons—will be given exoskeletons, like insects, to deflect bullets. These exoskeletons will have the capacity to turn into offensive weapons as well. These exoskeletons will also deflect light so that those wearing them will be as invisible as the man at the center of the Panopticon, as invisible as God. Ned Thomas, director of the Institute for Soldier Nanotechnologies at MIT, explains why he wants to try to create these übersoldiers—and I picture him laughing like all the

mad scientists in all the bad science fiction movies as he speaks—
"Imagine the psychological impact upon a foe when encountering
squads of seemingly invincible warriors, protected by armour and
endowed with superhuman capabilities, such as the ability to leap
over 20-foot walls."[1]

Military scientists long ago figured out how to put electronics into
the brains of rats, and to cause them to move forward, backward,
left, right by pushing buttons on computer keyboards. Imagine
the fun these scientists will have if they figure out how to do this
to women's hips.

Recent research has been aimed at co-opting the rats' will.
Scientists put an electrode near a pleasure center in the rat's brain,
and others to stimulate whiskers on each side of the rat's nose.
The scientists then trigger, for example, implants near the left
whiskers, and follow that by triggering the pleasure center. This
convinces the rat to move left. After only ten days of this, rats can
be trained to climb trees, walk, and stand in the open, or do many
other things rats don't normally like to do, controlled by technicians issuing commands from laptop computers up to 550 yards
away. As a reporter for the *Washington Post* put it, not disapprovingly, "The rat thus becomes a living robot, controlled remotely
by a human handler but able to go anywhere a rat can go."

"I like the results," said a scientist at Northwestern University,
who gave his reason: "This is the first time where you have control of a whole complex animal."

A scientist at New York's Downstate Medical Center put the
final word on this, "The rat looks normal and isn't feeling any
pain because he's getting rewards for doing the right thing."[2]

The rat is no longer a rat. It is a puppetrat, controlled by "providence," by God, by a man with a laptop.

Imagine putting electrodes near pleasure centers in human
brains. Imagine getting humans to feel pleasure for doing things

that are against their nature. Imagine getting them to feel pleasure for "doing the right thing," for doing that which is favored by providence, defined, of course, by those at the center. Imagine getting humans—or what used to be humans—to feel pleasure working for Wal-Mart (attaching RFID chips). Imagine getting them to feel pleasure purchasing items (containing RFID chips) from Wal-Mart. Imagine getting them to feel pleasure watching propaganda for the corporate state. Imagine getting them to feel pleasure voting in meaningless elections to put in power people who do not represent them. Imagine getting them to feel pleasure in following laws laid down seemingly not by those in power but by providence. Imagine getting them to feel pleasure as they narc out those who do not have implants or who otherwise do not choose to do "the right thing." Imagine getting them to feel pleasure in hunting down and killing those miscreants.

Imagine the fun these scientists will have when they put electrodes into the pleasure centers of women, to get them to feel pleasure—whether they feel pleasure or not—for "doing the right thing." They already do this: scientists have long since discovered that if they implant electrodes in women's brains—they use their patients in mental hospitals—they can bring the women, even women in what they describe as "a low mood," to have "repetitive orgasms."[3]

They may want to order a set of electrodes for use around the house.

Remote-controlled rats may be the least of our worries.

Tomorrow's warfare, according to experts at a conference on the future of weaponry, will be "revolutionised by computing, robotics and biotechnology to create 'killer insects' that can hunt down their prey in bunkers and caves and eat humans alive." Paul Hirst, professor of social theory at Birkbeck College, London University, gives some details: "micro aircraft that fly by their own sensors and carry many deadly sub-munitions; intelligent jumping mines that shower selected targets with small guided bomblets. . . . The result would be really effective substitutes for chemical and biological weapons: deadly bio-machines of finite life that could be released by sub-munitions, showering opponents in millions of nanobots . . . that could literally eat humans alive."[4]

And how will those in power find those they wish to have eaten? First, in addition to the RFID chips that can identify the location of someone who has bought any tagged consumer items, those in power will, according to Charles Heyman, editor of *Jane's World Armies,* be able to drop thousands of minimicrophones, cameras, and vibration sensors at crucial sites to relay information back to the center of the Panopticon.[5]

In case all of that doesn't suffice, military researchers are currently working hard to fabricate radar devices that will identify people by how they walk. It seems that our gait is as distinctive as our fingerprints, and scientists at Georgia Tech have been able to gain identification success rates of 80 to 95 percent.

A reporter asked Gene Greneker, head of gait research at Georgia Tech, whether he was concerned about the ends to which his work would be put. His response could have been spoken by the creators of mobile killing vans used by Nazis, creators of nuclear bombs, creators of electrodes to be put into the brains of rats (or women), creators of suits to turn the servants of those in power into übersoldiers. He said, "We are research and development people. We think about what's possible, not what the government will do with it. That's somebody else's job."[6]

The article did not report whether Greneker felt pleasure—electronically induced or otherwise—as he said this.

A couple of years ago, the United States government began bringing together information-gathering programs under a vast surveillance network called Total Information Awareness (TIA).

TIA was a program of the Information Awareness Office, which in turn is part of the Defense Advanced Research Projects Agency (DARPA), run by the Pentagon.

Those in charge would like to be able to provide their agents with instantaneous access to records from around the world. A lot of records. In its advice to corporations that may contract to provide some of this information, DARPA states, "The amount of data that will need to be stored and accessed will be unprecedented, measured in petabytes." One byte is the amount of memory it takes to store one letter. One petabyte is one quadrillion bytes. That's one with fifteen zeros after it. This means that those in power want to maintain a database that would be more than fifty times larger than all of the books in the Library of Congress, or somewhere on the order of a billion books.[7]

This information could include financial, health, shopping, telephone, employment, and library records, fingerprints, DNA samples, gait analyses, brain scans, surveillance photographs, information on whom and how you love (including audio and video recordings of your most intimate moments), recordings of phone conversations, copies of emails, maps of Internet activities, information on addictions or other exploitable weaknesses, and all sorts of other information no sane person could even dream of collecting. Even if the project were to use only one petabyte of storage, that would still be enough to amass forty pages of text for each person on the planet.

In response to criticism, the United States government changed the name of Total Information Awareness—though not, of course, its function—to the less accurate Terrorism Information Awareness. Presumably it also began dossiers on everybody who complained about the program.

The Information Awareness Office logo consists of the name of the organization surrounding a blue background against which we have the truncated pyramid and the by-now-familiar all-seeing eye. This eye, of providence, of God, of the police, of the military, of representatives of major corporations, emits a ray of golden light to illuminate and overlook the globe. In the upper right are the initials DARPA, and in the lower left is *Scientia est Potentia,* a Latin phrase they translate as *Knowledge is Power*.

Knowledge is not always power. There are other ways to be and perceive in the world. Knowledge can be love. It can be relationship. It can be connection. It can be neighborliness or familiarity. Knowledge can simply be knowledge.

Last week I had one of the most exciting and wonderful mornings of my life. I live near a pond. I often sit at its edge. I love to watch tadpoles swim, watch them over time grow legs, slowly lose tails, take their first hops onto land, make their first awkward flips of the tongue (sometimes before they learn how to use their tongues, they wildly miss their targets and their whole bodies tumble till they land on their noses!). I also love to watch whirligig beetles who skate in incomprehensibly complex patterns—or maybe in no patterns at all—over the surface of the water, and backswimmers who hang motionless then glide quickly toward potential prey. Newts who swim to the top for great gulps of air, then back down again too deep for me to see. I watch mating dragonflies, the male joining his genitals to the female's near the base of the female's head, leaving her back end free to dip into the water and drop eggs even as they mate.

That morning a large brown insect crawled from the pond, covered with mud. I'd seen insects like this, and I'd also seen their skins hanging empty from blades of grass. I didn't know who they became. So I watched.

I watched as the creature made its way slowly across spaces of bare ground and through patches of grass until it found the blade it wanted to climb. It made its way to near the top, then grabbed on tight.

I waited. I looked away to water skippers and willows and rushes. When I looked back a furry hump had formed on the creature's back, between where the shoulder blades would be on you or me. The hump got larger.

Again I waited. The wind played with the tips of redwood branches. Wrentits sang, as did sparrows and thrushes, and some other bird I could not name but whose trilling song made me smile. A jay cocked its head and looked at me.

The hump became a head, and over time first one, another, then a third pair of legs became visible. They were all the palest yellow, nearly white. They unfolded slowly.

I had no idea who this creature was. The sun rode the sky. It grew warm on my back. More of the creature emerged, and more. It began to hang from the shell that used to be its skin. Sometimes it would move vigorously, sometimes it would slowly expand, and sometimes it would rest. I wondered if it would keep pushing itself from its former skin until it fell to the ground. Then suddenly it thrust itself upward to grasp the grass with its legs. It pulled hard, and pulled again. Finally it was free.

I still had no idea who it was. It was pale and stubby, with ruffles on its back.

I wanted to take a picture to show my friends, to post on my Web site. But I knew, because the creature told me, that this would be wrong.

The ruffles on its back began to expand. Slowly. Everything was slow. I'd been sitting by then for probably two hours, but it seemed much less because each moment I wanted to know what would happen the next.

The ruffles unfolded, the abdomen expanded. Longer, longer. The ruffles became wings, four of them. The eyes clarified. Colors came alive.

It was a dragonfly. No longer pale pink but very bright blue. "Now," it said. "Now get the camera." I did. It spread its wings. I took pictures. It waited.

I was hungry. I walked the path—three-eighths of a mile through dense forest—to my mom's. As I walked I pondered how many times I've walked this path these past three years. Easily three to four thousand. For the first year or so I used to carry a lantern at night, but then I quit because I got to know the path well enough to walk it at a normal pace even on the darkest nights (hint: look up to see the slight break in the forest canopy that signals the path). This time, of course, it was early afternoon. I got to my mom's. I ate there. I often do. I made my own meal, but she often cooks for me. She likes to cook and knows how to do it well. She also knows what foods I like, or don't. Afterward I helped her in her garden. She tells me what chores she would like me to do, and I (eventually) get them done. It works. We each know what helps the other, and want to help the other how we can.

I walked back home, expecting the dragonfly to be gone. But it remained through the afternoon, and into the night.

I awoke around 9:30 the next morning. The first thing I did was go outside, expecting, again, to see only the husk of the dragonfly, clinging to the grass. But the dragonfly remained. I stopped a few feet away. It did not move. I looked down to my feet for just a moment—to make sure I wouldn't step on any baby frogs if I shifted my weight—and when I looked back up it was gone.

There was only one large dragonfly on the pond. It was bright blue. It circled, then rose up to fly around the meadow, then back down to the pond. Then back up, in wider and higher spirals till it felt it knew the landscape. Higher and higher it spiraled, until it flew over the top of the redwoods and into the world.

Knowledge, whether it is of a dragonfly, a path, my mother, me, a landscape, is not always power. There are other ways to be and perceive in the world. Knowledge can be love. It can be relationship. It can be connection. It can be neighborliness or familiarity. Knowledge can simply be knowledge.

. . . .

Or knowledge can be power over others.

Do you know as much about yourself as they do?

If you're a consumer, there are records about your credit cards, layaway plans, leases and rentals, purchases, purchase inquiries, subscriber lists, clothing sizes, Internet Web browsing preferences.

If you've been to school, there are records about your school applications, academic records, academic references, extracurricular activities, awards, and sanctions.

If you have a job, there may well be records about your job applications or employment agency applications, medical examinations, drug tests, personal and professional references, performance assessments, employment history, employment licenses and certifications.

If you indulge in any entertainment or leisure activities that involve spending money, there are records relating to your travel itineraries, recreational profiles compiled by travel agents, auto and other rentals and leases, lodging reservations, airplane, ship, and train reservations, concert or other entertainment tickets, newspaper and magazine subscriptions, and telephone/cable records.

Do you participate in financial transactions? Have you ever borrowed money or had a bank account? There are data on your bank records, ATM cards, credit card transactions, online banking files, credit reports, tax returns, stock brokerage accounts, and traveler's check transactions.

If you ever had an insurance policy, then you are assessed for the risk you pose to the company. If you have health, auto, home, business, or any other kind of insurance (and in this paranoid culture, you can buy insurance for just about any conceivable risk), then there's lots of information about you in those files.

If you've ever been an initiator or target of legal action, you might be mentioned in court records, lawyers' records, in arbitration or other out-of-court settlement records, or in newspaper reports about legal actions.

If you've ever bought or rented property, then you're a tenant or a mortgage holder and there's information related to your real estate purchase, sale, rental, or lease. (One of the first things the Zapatista rebels in Chiapas did when they took over towns was to pull the mortgage records from the courthouse and burn them, to much public applause.)

Are you now, have you ever been, or will you ever be the recipient of public assistance or private benefits? Chances are good you've got one or more of these: private pension records, social security records, health care records, records associated with employment benefits, unemployment benefits, workers' compensation claims, disability records, food stamps, veterans' benefits, or senior citizen benefits.

Do you use any public or private utilities? How about telephone, electricity, heating fuel, cable or satellite television, internet service, garbage collection, sewer, security services, or delivery services?

Are you a driver, a voter, a traveler? Have you ever been married or divorced? Were you born? There's a record on you.

Who collects all that information? Who stores it? Who's got access to it? Is it shared among agencies, among corporations?

Do the people who have this information have your best interests at heart? Do they really know who you are? What do they want from you?

When the people at the Information Awareness Office translate *Scientia est Potentia* as *Knowledge is Power,* they're not only defining knowledge very narrowly (and in a way my therapist friends would say "is highly diagnostic of their own personal issues and difficulties"), but they're also perpetuating another scam, one that most of us participate in more or less willingly, to our own detriment and to the detriment of the planet. This is that they have translated *scientia*—the root of the word *science*—as *knowledge*.

The spooks at the Information Awareness Office aren't the only ones who conflate science and knowledge. It's pretty common in our culture. I asked philosopher Stanley Aronowitz, author of *Science as Power,* among many other books, about this conflation. He said, "Science is founded on the idea that the results of its methods—which are very specific mathematical and experimental methods—are equivalent to what we mean by truth. The mythology holds that science describes physical reality, that science is truth. And if science is truth, instead of merely one form of truth, then all other forms of truth—all philosophical truth, all ethical truth, all emotional, spiritual, relational, experiential truths—are devalued. They're regarded as something else besides truth. Scientists may agree, for example, that there is something called artistic truth, but they—and I'm talking not so much about specific scientists (although this is often true) as I am about what the scientific worldview does to all of us—don't think artistic truth has anything to do with the material reality that the scientist investigates."

He continued, "Science is based on exclusion. And not just the exclusion of all these other forms of knowledge. It's full of exclusions. Logic, for example. In order to establish its authority it excludes what might be described as a critical logical analysis that derives not strictly from experiment, but from the less formal observation of any, say, philosopher or political or social theorist."

"Or human being," I added.

He said, "Scientists will say, 'That's all very interesting, but it's really got nothing to do with truth. It's just your opinion.'"

"Why do we care?"

"At the very least because if you can convince people that science has a monopoly on truth, you may be able to get them to believe also that the knowledge generated through science is independent of politics, history, social influences, cultural bias, and so on."

And in the bargain, you can get them to doubt their own experience.

This might be a good time to examine the etymology of the word *science*. It comes from the Latin *scientia,* from *sciens,* which means *having knowledge,* from the present participle of *scire,* meaning *to know,* probably—and here's where it gets exciting—akin to the Sanskrit *Chyati,* meaning *he cuts off,* and Latin *scindere, to split, cleave.* The dictionary tells me there's more at *shed* (presumably the verb, as in dog hair, not the noun, as in a shack).

So I look up *shed,* which derives from the Middle English for *divide, separate,* from Old English *scaeden,* akin to High German *skeiden, to separate,* which brings us back to our Latin friend *scindere,* and from there to the Greek *schizein, to split.*

We are all familiar of course with the root *schizein* because of its famous grandchild *schizophrenia* (literally *split mind*), which is a psychotic disorder characterized by a loss of contact with the environment, illogical patterns of thinking and acting, delusions and hallucinations, and a noticeable deterioration in the level of functioning in everyday life.

Science, scire, scindere, schizein, schizophrenia. A mind split into pieces.

It should come as no surprise, at least to etymologists as well as regular people with too much time on their hands, that the words *scientia,* translated to mean knowledge, and *science,* the main means by which people in this culture are presumed to gain this knowledge, have at their core the notion of splitting off, separating from. After all, the word *separate* comes from the Latin for *self, se,* meaning *on one's own* (which springs from the belief and

promotes the fiction that a self is independent of family, community, landbase), and *parare,* "to prepare." In this culture it is separation that prepares a person for selfhood. It is separation that
defines us. Separation has become *who we are.* It is the illusion of
separation, as we shall see, that keeps us enslaved.

Surveillance, and this is true for science as well—indeed, this is
true for the entire culture, of which surveillance and science are
just two holographic parts—is based on unequal relationships.
Surveillance—and science—requires a watcher and a watched, a
controller and a controlled, one who has the right to surveil or
observe—with knowledge, truth, providence, and most of all
might on his side—and one who is there for the other to gain
knowledge—as power—about.

These unequal relationships require a split, a separation.
There can be no real mixing of categories, of participants. The
lines between watcher and watched, controller and controlled,
must be sharp and inviolable. Humans on one side, nonhumans
on the other. Men on one side, women on the other. Those in
power on one side, the rest of us on the other. Guards on one side,
prisoners on the other. At Pelican Bay State Prison, where I
taught creative writing for several years, I once received a chiding
letter from my supervisor after I innocently answered an inmate's
friendly question as to what I was doing for Thanksgiving: to
even let him know I was spending it with my mom was to make
myself too known—too visible—to this other who must always
be kept at a distance.

If this sounds a lot like the pornographic relationship, that's
because it is. Pornography—cousin to surveillance, and bastard
child of science—requires the same dynamic of watcher and
watched, the same dyad of unchanged subject gazing at an object
to be explored at an emotional distance, the same relationship of
powerful viewer looking at powerless object. (This may explain

at least some of the popularity of pornography: people who are powerless in every other aspect of their lives get to feel some power as they look at these pictures and read the attached text.) When I read that we must not "make scruple of entering and penetrating into these holes and corners," I wonder whether I am reading a letter by the father of science Sir Francis Bacon to King James I (describing how the methods of interrogating witches— that is, restraint and torture—must be applied to the natural world), or whether I'm reading a description at www.perfectly-pussy.com. When I read about using the "mechanical arts" (that is, once again, restraint and torture) so that she "betrays her secrets more fully . . . than when in enjoyment of her natural liberty," am I still reading Bacon's words on science, or have I landed at www.fetishhotel.com?[8]

These unequal relationships—insofar as we can even call them relationships—must be oppositional. Predator and prey must not be working together for the benefit of both of their communities, and for the benefit of the land. Instead, from this perspective— this perspective based on selves being separate, and knowledge being gained through splitting off—predator and prey (and this applies to humans as well) must be locked in an eternal battle, good against evil, a battle that ends in Armageddon.

As civilization plays out its grim endgame, and as those in power move ever closer to their ultimately unattainable goal of absolute control (through absolute surveillance), converting in their efforts the wild both inside and out to devastated psyches and landscapes, it might be well past time to reconsider the premises that underlie much of this destructive way of being (or not being) and perceiving (or not perceiving). For in many ways, perception shores up the whole bloody farce.

So often we see the world—or do not see the world, but see what we project into the world—in terms of opposites. Given that our culture is based on opposition, this is precisely what we should expect. Any discussion of surveillance highlights many pairs of seemingly opposing—warring—impulses.

There is the need to control versus the need for freedom. The need for bureaucracies to run large institutions versus the need for democracy. The need for administration and regulation of markets versus the need for markets that are free and self-balancing. The needs of industrial production versus the needs of our landbases. The requirements of scientific analysis versus love and relationship. The needs of technology—with each new technology separating us further from the natural world—versus our need to be intimate with nature. The needs of efficiency versus the requirements (and joys) of craft. The needs of police forces versus the needs of people acting in self-defense. Paternalistic governance versus autonomous humans in communities.

In the wars between these perspectives, right now the winners in each case are those who are served by corporate and governmental bureaucracies.

Some of these opposites really are opposites. Industrial production really does destroy landbases. The *need* for control really does run counter to freedom. Scientific analysis cannot coexist with love and relationship (vivisection, anyone?). Industrial technologies inescapably separate us from nature (that's the point!). And so on.

In these truly oppositional cases, we are not facing the actual conflict. We could not, and continue to live as we do. No reasonable people would destroy their landbase, even to manufacture groovy products like G.I. Joes and Hummers (there are now more automobiles than people in the United States, by the way). Similarly, no one not gripped by fear would require—or allow—security to throttle freedom. (Are industrialized people free? As we'll see in a moment, we are "given" the freedom to make false choices, the freedom to choose anything we want so long as it does not go against the structures of power.)

Instead these wars are fought on the landscapes of our minds.[9] Certainty versus mystery. Logic versus emotion. Security versus freedom. Personal consumption versus service to community.

In these wars, fear and habit have been defeating courage and freedom.

A classic device of power—and this is true whether we're talking about emperors or perpetrators of domestic violence—is to present their victims with a series of false choices whereby no matter which the victims choose, the perpetrators win and the victims are further victimized. Nazis, for example, sometimes gave Jews the choice of different colored identity papers. Many Jews then focused, reasonably enough, on trying to figure out which of these colors would more likely save their lives. Of course the color of the identity papers made no material difference: the primary purpose of the choice was to divert victims' attention from the task of unmaking the whole system that was killing them. In addition, this false choice co-opted victims into believing they were making meaningful choices. In other words, it got them on some level to take responsibility for what was being done with them: *If I am killed it is my own fault because I chose the wrong color.*

Now, would you rather vote Republican or Democrat? For which major corporation would you like to work? Which shopping mall has the best deals this weekend? Do you want privacy or security?

Both the spies at the Pentagon—or maybe at some corporate headquarters, we always get confused as to which is which—and public interest advocates at, for example, the American Civil Liberties Union (ACLU) state that we should and must have security *and* privacy. But it's another false choice, both sides of which, framed as they currently are, serve to divert our attention while those in power continue to extend their control. The military industrial complex continues to operate as if spying on "our enemies" will make each of us individually—separately—and the world in general more secure. And lawyers continue to operate as if more regulations will stem the tidal wave of invasive technology and commercialization of culture. Neither of these positions has a shred of evidence to support it. In fact both are demonstrably untrue. Nonetheless they are clung to, articles of faith in some religion to which we do not—cannot—admit we adhere.

Not only the spies and übersoldiers are invisible. So are our beliefs.

Aronowitz told me that a "fundamental precept of science is that at some point we're finally going to get to the bottom of things, that we are getting closer and closer to 'the truth.' That someday we'll understand the fundamental building blocks of matter, we'll unify electromagnetism with gravity, and, like Einstein believed, we'll have a general theory. This conceit may be scientists' version of utopian hope. Now, some scientists don't believe this, instead believing there will always be uncertainty and upheaval, but the majority believe we're moving toward some ultimate theory. And you really have to say that this latter group subscribes to science as a religion."

"What's wrong with that?" I asked.

"First, their religion masquerades as something it's not; in fact they consider themselves skeptics, and are often highly scornful of people who rely on 'mere faith.' More broadly, science is coercive in the same way that dogmatic belief in a deity can be coercive. Just as God is then taken as an axiom by true believers, so the four elements of scientific discourse cannot be questioned."

"Four elements?"

"The first is the exclusion of the qualitative in favor of the quantitative. If you cannot assign a number to something, it doesn't exist. The second is that except at the outset, speculation is excluded in favor of observation and experimentation. The third is that knowledge is claimed to be free of value. There's nothing inherently wrong in knowing how to make a neutron bomb, for example."

"Those who make them can say, 'We are research and development people. We think about what's possible, not what the government will do with it. That's somebody else's job.'"

"It's simply information, so the mythology goes. And the fourth is that method is given primacy in the confirmation of knowledge."

"Meaning . . ."

"Meaning that since science has defined its methods as the only way to discover truth, the only acceptable criticisms of science are those conducted within the methodological framework that science has set up for itself. Further, science insists that only those who have been inducted into its community, through means of training and credentials, are qualified to make these criticisms.

"Many scientists consider it absurd that Christian fundamentalists use biblical references to bolster their claim that the Bible is literally true, yet we all let science use the tools of science to settle disputations concerning its own viability. That makes no sense to me.

"But there is something else at stake here. Theirs is a belief in the end of history. It's a version of a belief, at the level of science, of what Francis Fukuyama says at the level of human affairs, that we've finally ended history.[10] Fukuyama thinks we've ended history because the world has been unified under the common denominator of capitalism, so-called liberal democracy, the market."

"Obviously there are a lot of problems with Fukuyama's boosterism of capitalism, not the least of which is that it doesn't match reality. . . ."

"Oh, absolutely. He makes no sense at all. And the same is true scientifically. The scientific hope is for an end of natural history. We will someday understand everything."

"And essentially be as God. . . ."

"But let me ask you this: Does the world change? Is the material world itself moving constantly?"

"Absolutely."

"Well, if that is true, then we can't ever know the ultimate anything. If there exists anything even remotely resembling freewill actors anywhere in the universe, then there can be no ultimate knowledge of the sort science purports."

Control

No matter how much lip service those committed to power may pay to the principle of equality, they can never approach their fellow human beings on an equal footing; their relationships with others are defined solely in terms of power or weakness. Therefore, they must accumulate as much power as possible, with the aim of becoming invulnerable and proving this invulnerability.

Arno Gruen

You're in a car, and you get pulled over by a cop. Maybe it's a soldier. Maybe it's a Starbucks agent. You can't tell. You roll down your window to see your own reflection in the shiny sunglasses.

You start to protest that you were driving only four miles over the speed limit, and your speedometer isn't accurate, and you were on your way to get it fixed, when the cop silences you by looking over your left shoulder into the back seat.

The cop says, "Where did you get those groceries?"

"Safeway," you say. You wonder if the cop works for Albertson's.

"Where were you yesterday?"

"What? I—?"

"It says here that you left your home at 7:45 and didn't make it to work until 11:15."

"Says where?" you ask, more incredulous than outraged. Fortunately the cop doesn't notice that you, not the cop, asked a question. This was a breach that could have cost you dearly. Everyone knows that those in authority ask the questions, and that those not in authority answer them. Isn't that what you learned in school?

You notice that the cop is talking. "You stopped as you always do for a cup of coffee." The cop pauses before saying, "Not at Starbucks."

You start to say that you're trying to quit drinking coffee altogether, but that in the meantime you prefer not to support that chain. Then you remember that anything you say can and will be held against you in a court of law, or in a marketing plan.

"And then you went somewhere else, didn't you?"

Your face remains impassive.

"You went downtown."

Then suddenly you remember the RFID chips. They're in your clothes. They're in your car. They're everywhere. How could you have been so stupid?

Okay, let's tell three versions of what happens next.

Version one: The cop, or maybe the Starbucks agent, says, "Someone committed acts of terrorism this morning."

"I don't know what you're talking about."

"I think you do. Several people hacked into the Starbucks computers using a wireless connection from somewhere close to the store. The machines ceased to work."

"No one was hurt," you say, before you can help yourself.

"People were unable to purchase their frappacinos. Commerce was impeded. To damage a machine and to impede commerce is to commit an act of terrorism."

At least the cop doesn't know who I was with, you think.

"And we know who you were with. We know what time the other people arrived, and we know where they went afterward. Step out of the car, and make sure to keep your hands where I can see them."

Version two: The cop, or maybe the Starbucks agent, says, "Someone committed acts of terrorism this morning."

"I don't know what you're talking about."

"I think you do. Leaflets were handed out urging people not to patronize McDonald's, Burger King, and . . ." You can tell the cop is enjoying this. "And Starbucks."

"No one was hurt," you say, before you can help yourself.

"Commerce was impeded. To impede commerce is to commit an act of terrorism."

At least the cop doesn't know who I was with, you think.

"And we know who you were with. We know what time the other people arrived, and we know where they went afterward. Step out of the car, and make sure to keep your hands where I can see them."

Version three: The cop, or maybe the Starbucks agent, says, "Someone committed acts of terrorism this morning."

"I don't know what you're talking about."

"I think you do. There was a protest yesterday. There were acts of terrorism. Leaflets were passed out urging people to buy less. And windows were broken at McDonald's, Burger King, and . . ." You can tell the cop is enjoying this. "And Starbucks."

"I didn't know anything about that. My resolve broke this morning, and I just had to have a Starbucks caramel machiatto. Nothing else would do."

The cop looks at you, face unreadable.

"Honest, I do love Starbucks, I swear, and I was just buying coffee when these protestors rushed in. Nasty, awful protesters."

"You say that now," the cop says, "but at the time you did not step in. You failed to assist McDonald's, Burger King, or even . . . Starbucks."

"I didn't do anything."

"That's the point. You didn't help out. Step out of the car, and make sure to keep your hands where I can see them."

What does it take, in your mind, to be guilty of terrorism? Of disloyalty?

• • • •

Right now the Pentagon is developing an urban surveillance system that will be capable of tracking, recording, and analyzing the movement of every vehicle in a city. The system will use computers and thousands of cameras. Face-recognition software will identify not only vehicles but drivers and passengers. The Pentagon's name for this project: "Combat Zones That See."

The point, as military analyst John Pike puts it, is that the "government would have a reasonably good idea of where everyone is most of the time."

And it will not only be the military. Pike says that once the Pentagon "demonstrates it can be done, a number of companies would likely develop their own version in hope of getting contracts from local police, nuclear plant security, shopping centers, even people looking for deadbeat dads."[1]

In related news, all students in all public schools in Biloxi, Mississippi will now be monitored at all times. Five hundred cameras have been installed throughout the school district, in every classroom and hall. Once they catch some kids smoking in the boys' room, we can presume bathroom cameras will follow, as well as, of course, cameras in the girls' locker rooms (which could be a source of revenue to offset the $2 million it cost to install the system). School administrators will view the cameras on the Internet.[2]

The Enterprise Charter School in Buffalo, New York has gone one step further. All 450 students are required to wear plastic identification cards around their necks. These cards contain their photographs, names, and grade levels. They also hold embedded RFID chips. Every time a child enters the school, the child is required to approach a station where an electronic receiver interprets signals from the chips. The goal, according to school administrators, will be to monitor everything from library loans to disci-

plinary records to cafeteria purchases to the students' every movement, tracking, for example, precisely what time each student
arrives in each class, and precisely what time and at what location
each student gets on or off the school bus. To keep kids from just
handing their cards to other students while they cut class, the
RFID receiving stations are also monitored by video cameras.

It may or may not be significant that the tags the students are
forced to wear contain the same type of chips as those worn by
inmates at the Pima County Jail in Texas.

But I guess we shouldn't view the students as prisoners: the students get to wear the ID chips around their necks, while the
inmates have to wear them on their wrists. So there *is* a difference.[3]

In Iraq, U.S. soldiers are herding old men and young boys into
barbed wire enclosures, holding them hooded and handcuffed in
the heat until they determine whether or not they are, to use the
words of an American major, "bad guys." Bad guys, that is, those
who do not do "the right thing," are imprisoned indefinitely.
Those who are not "bad guys," that is, those the soldiers do not
perceive as opposing them, and therefore opposing the desires of
those in power half a world away, are released. Till next time.

The major continued, in language that captures perhaps as
well as Jeremy Bentham ever could the essence of surveillance (I
was going to write coercive surveillance but remembered that all
surveillance relies on that imbalance of power, and is therefore
coercive) and of large-scale social control. "What we are doing
here may seem harsh," he said. "But we explain to them that to
catch a single fish you have to cast a big net. They understand if
they have nothing to hide, we will release them."[4]

In South Carolina in 2003, U.S. police herded American teenagers
against the walls of their high school, handcuffed them, and held
them until they could determine if they, too, were bad guys.

Before moving into the school, police had spent several days monitoring the school's video cameras, and said they saw what they perceived as suspicious activities, including students "concealing themselves from the cameras."

Having seen these suspicious activities, the police (along with the school administrators, insofar as there is a fundamental difference) evidently decided to cast a wide net. Fourteen police went to the school, and in the words of one cop, "assumed strategic positions," after which they moved to "safely secure the 107 students who were in that hallway. During that time some of the officers did unholster in a down-ready position, so that they would be able to respond if the situation became violent."

Here's how that action was described by a fourteen-year-old student: "They would go put a gun up to them, push them against the wall, take their book bags and search them. They just came up and got my friend, not even saying anything or what was going to happen. . . . I was scared."

Another student, however, had a different response. He was "kind of nervous," he said, but not scared, "because I didn't have anything to hide."[5]

The stated reason for the raid was to halt drug sales. Although students reported that the identities of those who did sell drugs was widely known, the police did not arrest those people but chose instead to stage a dramatic raid, during which no drugs were found. But was busting drug dealers the real purpose of the raid? If a central purpose of fingerprinting children in lunch lines, forcing children to wear RFID tags, and putting video cameras in schools is to normalize surveillance in the lives of these children, so, too, this raid can serve to normalize police violence, which, from the perspective of those in power, is nearly always a good thing.

Scientists map human genetics. Medical engineers create new synthetic materials to replace aging parts of your body. Molecular biologists and computer engineers work together to meld living, reproducing organisms with silicon sensor chips. Geneticists create an embryo that includes both human and rabbit DNA, and add frog genes to strawberries. Engineers probe the molecular and atomic structure of the universe, as scientists lay waste along the boundaries between knowledge and mystery.

Most of the news stories of these events are written by professional journalists. Their charge is to write "objective" articles that announce scientific breakthroughs without revealing any bias for or against the experiments. Then professional ethics journalists write articles that cover the ethical controversies without taking one side or another. But objectivity does not exist. Everyone has a perspective. Everyone is a subject. Objectivity is a cover for support of the status quo.

In this case, the status quo is full-steam-ahead technocracy, the turnover of ethics and governance to the unholy coalition of technical knowledge and wealth-based power. Edward Abbey got right to the point when he wrote that most of us "suffer not merely from vulgar, childish incredulity in regard to science-and-technology, but from a serious streak of moral servility. The majority gape at the activity of our scientific technicians in the attitude of awestruck genuflection. On your knees, commoners! . . . Meanwhile, the obvious, inescapable, simple truth of American life is that we are surrounded by gibbering electronic machines that make daily existence a neurotic ordeal, while overhead, undergoing constant refinement, hangs the nightmare of total annihilation. . . . Recent history demonstrates that scientists, as a class, are as happy working for a Hitler or Stalin as for the oligarchs of the USA. Most scientists work their specialized niches as servants for commerce, government, or war. (And what's the difference?) These millions of little men in white smocks, busily torturing atoms and small animals in their laboratories, are generally up to no good and should be kept under strict public control."[6]

. . . .

Aronowitz continued, "For some scientists everything outside the box—defined by the rules of scientific discourse—must be ignored. And they often get very agitated when you call them on the game they're playing."

"And the game is . . ."

"Religion. Teleology. Control. The desire for prediction, and ultimately the desire to control the natural world, has become the foundation of their methodology of knowing truth.

"Think about it. I mean, what is a laboratory experiment? At the beginning one must select, from the multiplicity of objects and relations that constitute the world, a slice to study. How do you conduct a laboratory experiment? The first thing you do is factor out the world. You factor out emotion. You factor out ethics. You factor out nature, if you want to put it that way. You factor out the cosmos. You create a situation of strict abstraction. From that, we think we can extrapolate propositions that correspond to the world and its phenomena. Or rather scientists think that. And these propositions do correspond to the world, so long as we ignore the actual physical world and its context."

I thought about how the first law of science (as well as of bureaucracy) is to know the laws and rules. Under this way of thinking, or rather not thinking, the realities of particular people—whether these people are humans or nonhumans—in particular places in particular circumstances are secondary, if not irrelevant. It's all about the rule of law, no matter which fundamentalist religion we're talking about: Christianity, Science, Capitalism, Progress. And the way to get people to follow these laws (in addition to the direct or indirect threat of violence, of course) is to get them to disbelieve in their own direct experience and to believe instead in the "objective" observations of an elite priesthood, whether this priesthood is, once again, Christian, Scientific, Capitalist, Bureaucratic, what have you. The observations of the priesthood will be mediated by technology and will follow a model (that is based on the "observations" of this priest-hood, using their proprietary technologies). When you come to see this, you begin to cut through the culturally induced miasma to remember that personal observation—which *must be* disdained by science and scientists (as well as other types of priests)—is primary, and scientific (religious, economic, bureaucratic, etc.) "observation" is in fact speculative.

I asked Aronowitz, "What are the social implications of this?"

"The point of science—and this may or may not be true of individual scientists—is to make the world subject to human domination. If they can abstract, and then they can predict on the basis of that abstraction, then they can try, at both the human and natural levels, to use that prediction in order to exert control.

"Genetic engineering is a great example, although almost any field will work as well. The ideology underlying its conceptualization is that we cannot and will not depend on nature to yield its own productivity, both in terms of its own development and human need. We're going to intervene, because the process of maturation has to be faster, the output has to be more plentiful, production has to be cheaper, humans have to be more in control of the process."

Genetic engineering is useful for other things as well.

As the people at the Sunshine Project—an organization dedicated to halting the use of biological weapons—put it, "Conventional and current wisdom holds that population specific biological weapons are practically and theoretically impossible. At a practical level, many scientists consider it impossibly difficult to use genetic variability to create a weapon capable of killing or otherwise harming a specific population. Other scientists, including many geneticists, argue that no suitable 'ethnic genes' exist in the first place. Both notions are wrong." They continue, "[A] recent analysis of human genome data in public databases revealed that hundreds, possibly thousands, of target sequences for ethnically-specific weapons do indeed exist. It appears that ethnically-specific biological weapons may well become possible in the near future."[7]

Right now, for example, little men in white smocks are searching for specific genetic variations between U.S. residents with European ancestry, Han Chinese, Japanese, and Yoruba people of Nigeria. They are doing this to the tune of $100 million (much of that public money). Further, the U.S. National Institute of Justice has committed to determining "race proportions" from crime-scene DNA. Work has been done in India to attempt to determine the genetic markers that may differentiate castes, and in Spain to determine genetic markers differentiating Basques from non-Basques.[8]

Just to make sure none of us think that those in the white smocks and most especially their masters want to discern these differences merely out of some sense of burning curiosity, we should consider these words from the Project for the New American Century, an organization boasting many members of the current U.S. administration and with a philosophy that has to this point predicted and indeed driven much U.S. foreign and domestic policy: "[A]dvanced forms of biological warfare that can 'target' specific genotypes may transform biological warfare from the realm of terror to a politically useful tool."[9]

We can recoil in horror and mutter impotent demands that someone oughta do something about all this, but these are *our* scientists and engineers. They arise from our culture, and most of us—at least most of us industrialized humans—most of the time share their worldviews, lifestyles, and ethics. We also share in the fruits of their work (we are explicitly excluding the poor and nonhumans from this, as do those in power, and that's the point).

Who is "the public," and how can it control those millions of scientists, those millions of little men?

The "public" in this culture has been fragmented into a collection of functional sectors: farmers, scientists, factory workers, middle-level corporate managers, policemen, legislative aides, mall shoppers, public relations consultants, radical forest activists, and so on. It is not and has never been (once again, within this culture) a nonhierarchical community composed of equal voices with decision-making power. The global economy trumps local subsistence ecology. Science trumps place- and myth-based spirituality. Money trumps common sense as well as ethics. Science pushes us to try to know everything. The earth is losing its wild places and wild people. Our everyday decision making is controlled by rational thinking (or rather the rational thinking of the schizophrenic, except oftentimes diagnosed schizophrenics make more sense and display a larger truth[10]). Technological research is driven by the military/security apparatus, and public universities are dominated by military-related funding. National security is trumpeted as a criterion for relating to the peoples of the world. Communities have no say in what factories will be built in their town squares, or what chemicals will be dumped into their drinking water. This isn't science fiction. This is your world.

Given all this, how do human beings, if there are any left, get stuck, for example, in a confrontation between their ostensible needs for security and privacy? Do security or privacy actually correspond to any past or current reality?

Is the problem spying? Is the problem spying by those with power? Is the problem technologies that allow or facilitate spying? Or is it something else?

Is information power? Is secrecy power? Is technology power? What is power?

How do you want to live?

. . . .

Let's take a quick multiple-choice test.

Which of the following do you fear the most?

a. Big black men who are the great-grandchildren of slaves
b. Crewcut men with funny black radio transmitters
c. Car salesmen, junk emailers, and people who call you at dinnertime to try to sell you vacation timeshares
d. Planned community architects
e. Global warming
f. The fact that half the U.S. drinking water supply is contaminated with carcinogens
g. The worldwide biodiversity crisis
h. Cops
i. Übersoldiers
j. Terrorists, a.k.a. people who fly airplanes into buildings
k. Terrorists, a.k.a. people who fly military planes over buildings and drop bombs
l. Terrorists, a.k.a. CEOs of major corporations
m. The poor
n. The natural world
o. Industrial civilization
p. Living without industrial civilization

Which of the following do you prefer as a way to deal with social problems?

a. Writing to your so-called political representative
b. Ignoring problems and assuming the techies will iron them out
c. Self-medicating with tobacco, alcohol, and shopping
d. Getting the best deal at the mall, and avoiding taxes and fees
e. Bombs
f. Processing your fears away by chatting with friends
g. Cops, übersoldiers, and the largest, most expensive military in the world
h. What problems? I don't see any problems.

These are not rhetorical questions or straw men. Our fears and our responses are the moment-by-moment choices that determine the course and quality of our lives.

When earlier we described the military's attempts to create über-soldiers, we're afraid we sold the military a bit short. Those at the Pentagon want to do more than create invisible exoskeleton-clad soldiers who can leap tall buildings (capes optional). I have before me a postcard inviting me to the "DARPA Bio Booth" at a big DARPA conference, which is called "Harvesting Biology for Defense Technology." The focus of the first day—and dust off your old books of feminist theory before you check out their terminology, because the feminists were right, those military wonks really do get off on this stuff—is "Enhancing Human Performance" ("Thrust areas include: Neurosciences; Metabolic Engineering; Nutrition; Genomics"). The focus of the second day is "Protecting Living Assets"—by which they mean übersoldiers, not living landbases—("Thrust areas include: Environmental Detection and Surveillance; Biomedical Science Technologies; Biological Processing and Manufacturing"). The third day they'll study "Improving System Performance" ("Thrust areas include: Biomimetic Materials, Assembly, Manufacturing; Biomimetic Robots; Intelligent Machines; Biomimetic Signal Processing and Sensors").

At the end of the third day they will see everything that they have made, and behold, it will be very good. And on the fourth day they will rest.

Before we describe in more detail some of their "thrust areas," let's take a look at some of their workshops. Monday morning they start off bright and early with "Metabolic Engineering & Dominance," followed not by "Leather & Submission" but by "Persistence in Combat," then "Continuous Assisted Performance," "Brain Machine Interface," "Augmented Cognition" (can we hope for a guest demonstration involving the Commander-in-Chief?), and so on. The most interesting workshop on Tuesday will be "Triangulation Identification for Genetic Evaluation of Risks (TIGER)." Now, I don't know about you, but when I hear those in power talk about things like "Identification for Genetic Evaluation of Risks," I immediately think of four things. The first is smallpox-laden blankets. The second is the Tuskegee Syphillis Study, in which several hundred black men were recruited to participate in what they thought was a study of treatments for "bad blood," but what was in reality titled "A Study of Untreated Syphilis in the Negro Male." For the next forty years, until the study was halted, findings were regularly published in medical journals and in public reports.[11] The third is that the United States is currently planning to build at least three new bioweapons laboratories dedicated to the creation of new classes of toxins, including genetically engineered toxins. Presumably the mantra of those who work there will be: "We are research and development people. We think about what's possible, not what the government will do with it. That's somebody else's job." If they repeat this often enough they may eventually believe it. Or at least they may believe that they believe it. But even if they do, that will not protect them from someday following their moral predecessors to their own Nurembergs. The fourth thing I think of is the line we mentioned a few pages ago, from a document called *Rebuilding America's Defenses* put out by the Project for the New American Century, an organization with members including Vice President Dick Cheney, Secretary of Defense Donald Rumsfeld, Jeb Bush, and Paul Wolfowitz, generally considered the "mastermind" behind the invasion of Iraq. On page sixty, the authors state, if you recall, that "advanced forms of biological warfare that can 'target' specific genotypes may transform biological warfare from the realm of terror to a politically useful tool."[12] These people, too, will face their own versions of Nuremberg.

Other exciting workshops for the day include Biosensor Technologies and Activity Detection Technologies. Day three starts with a rush: Biologically Inspired Multifunctional Dynamic Robots. They move from there to Controlled Biological Systems (which I'm presuming is a description of their desired endpoint for the entire planet). Later in the day they have Spectroscopic Observation of Remote Environments, and they finish with Exoskeletons.[13]

We have mentioned before the Brain Interface Program, in which rats are convinced by means of electrodes in their brains to do "the right thing." But we didn't mention that the chief of the project, Alan S. Rudolph, has what's left of his heart set on transmitting images and sound directly into rats' brains. That would, of course, just be a first step. Later will come soldiers, and perhaps more interesting, prisoners: the potential for mind control (and torture) is staggering. Probably literally. But certainly the most interesting use of all would be to transmit images and sound directly into the brains of the general public. Oh, we forgot, those in power already have that capacity: it's called television. But at least we can turn off (or smash) TVs.[14] Imagine the agony of living your entire life with advertisements for Coors Light, McDonald's, and the U.S. Army blasted continually and directly into your brain, and you can never turn them off. (Wanna take bets on how many years before this happens? My bet is ten. And I have to tell you that when the time comes that these ads are blasted directly into my brain I'm going to buy a gun. And lots of bullets. And when that time comes, I will, as the cliché goes, save the last bullet for myself. But only the last one.)

The next workshop is on Persistence in Combat. Let us translate this militarese into English. When soldiers are wounded, they are often rendered ineffective as fighting "assets." Technologies aimed at Persistence in Combat would have the goal of reducing pain or stopping blood flow. The primary purpose—indeed the sole stated purpose—is not to save the lives of young men and women, nor even to alleviate their suffering, but to allow them to persist in combat, to keep fighting (that is, killing) despite broken legs, despite having been shot in the gut.

Riddle: What do you get when you combine the Brain Interface Program with Persistence in Combat?

Answer: Electrodes in soldiers' brains that disallow them from feeling any pain whatsoever.[15]

Combine those electrodes with pills being developed that inhibit fear ("mastering the means," one journalist wrote, "of short-circuiting the very wiring of primal fear") and those in power have taken another step toward destroying their soldiers' humanity (that is, what remained after boot camp). The last of this humanity can then be made to disappear forever whenever a soldier pops a pill—also being developed—that can, to quote the same journalist, "over the course of two weeks, immunize him against a lifetime of crushing remorse," that is, a pill that medicates away one's conscience. (Dr. Leon Kass, chair of the President's Council on Bioethics, calls this "the morning-after pill for just about anything that produces regret, remorse, pain, or guilt.")[16]

One final workshop should satisfy your taste for the conference. Continuous Assisted Performance has nothing to do, surprisingly enough, with Viagra. It has to do with attempts to find biotechnological means to allow soldiers to persist in combat for up to seven days without sleeping. This is, as you may know, nothing new. Amphetamine use by soldiers—legal, enforced—is routine and has been at least since World War II. It's used to keep them awake. (And it has the added benefit of reducing pesky premature ejaculation.[17] Now *that* should make the wonks happy.)

In case those in power have not been clear enough about the level of control they aim to assert, we should consider the words of a 1996 report by the U.S. Air Force Scientific Advisory Board, which speaks of "the [hoped for] development of electromagnetic energy sources, the output of which can be pulsed, shaped, and focused, that can couple with the human body in a fashion that will allow one [and we know who this "one" is] to prevent voluntary muscular movements, control emotions (and thus actions), produce sleep, transmit suggestions, interfere with both short-term and long-term memory, produce an experience set, and delete an experience set. This will open the door for the development of some novel capabilities that can be used in armed conflict, in terrorist-hostage situations, and in training."[18]

Are you scared yet?

Let me ask this another way. Whom do these übersoldiers and übercops serve?

Let me ask this yet another way. Do science, technology, and the military better serve living breathing human beings, or corporations? Hint: When was the last time you saw armored police officers fire tear gas at corporate CEOs for deciding to allow cancer-causing chemicals in the workplaces of millions of Americans?

Do you believe this society's power structures have been designed in your best interests? The interests of your family? Your community? Your landbase?

What do these structures protect?

How do you want to live?

Identity

The price that must be paid for that kind of [oppres-
sive] violence is a constant nagging suspicion that,
deep inside, one is living a lie.

Arno Gruen

The system could not perpetuate itself if we were able to see it
clearly, or even if we allowed ourselves to think about it clearly. If
we did, we would not perpetuate it. We could not. But instead
our attention must be constantly diverted from real problems to
those that parasitize, trivialize, and make diffuse the outrage we
would otherwise feel. If these parasitic problems are made to
resemble the real problems, they keep people from ever looking
at the system itself. All of this is as true for dysfunctional families
as it is for dysfunctional cultures.

Here's an example of this sort of distraction. George spent a
couple of summers working in a salmon-canning factory in
Alaska. "Health regulations" required workers, including
George, to cut their long hair. All of the hippies he worked with
were outraged at this violation of their personal freedom, com-
plaining bitterly about this as they worked long hours at these difficult, tedious, dangerous jobs. Workers were losing their fingers
to the knives and gears of the canning machinery but worried
about their hair. To be forced by a company to cut their hair was
seen as an outrage, yet to be forced by an entire economic and cultural system to work a job they didn't love (or, for that matter, to
work a job at all) went as unnoticed as the hours that slip away in
one's so-short life. The outrage that might have been aimed
toward the whole system was (and is) diverted toward a vastly
smaller, less threatening problem.

All of which brings us in a roundabout way to "identity theft." Anybody who's read a newspaper or seen a television these last few years has probably heard the horror stories. Someone obtains identifying information about you such as your social security or credit card numbers, and begins to impersonate you (at least financially), then to steal your (financial) identity. The next thing you know, your bank account is empty, your credit card is maxed out, your credit rating is down the toilet, and you may be wanted by the cops for passing bad checks. That's when the bill collectors start calling at seven in the morning.

Here's a case study. Three years ago Philip Cummings spent three months at a desk job at Teledata Communications, a small software company on Long Island. During that time he was able, according to an MSNBC analyst, to "raid the nation's entire credit reporting system. And in the process, if the charges prove true, he could have sold virtually any American's digital identity." Before his ultimate arrest, he had sold the credit reports of thirty thousand people, including bank account information, credit card numbers, and former and current addresses. As the analyst notes, "It's always been easy to steal credit card receipts from restaurant customers or garbage cans. But thanks to the Internet, criminals can steal them by the thousands now; and more importantly, they can sell them by the thousands to other criminals. Nearly every Internet-based crime, from auction fraud to child pornography to stalking, starts with an identity theft. . . . [W]e now know the crime is so common that complete digital dossiers sell for only $30 a pop on the black market."[1]

Do you want to know Attorney General John Ashcroft's social security number? How about that of CIA Director George Tenet? How about their home addresses? You can get all those for $26 each. To get someone's bank balance costs a little more, about $300. As privacy advocates have been pointing out for a long time, just about all of this information is available for sale on the Internet.[2]

According to the Federal Trade Commission (FTC), ten million Americans have had their (financial) identity stolen in the past year, and twenty-seven million in the last five years. Identity theft cost the victims $5 billion and various businesses $48 billion.[3] The FTC's study reports that two thirds of the victims had their credit cards misused, 19 percent had their bank accounts pilfered, and 12 percent had charges placed on their telephone or Internet accounts.

Half the victims don't know how their identification was stolen; the rest knew it had been physicially lost or stolen from their wallet or mail, or during a financial transaction involving their account numbers. Twenty-six percent of the victims knew who the thief was, and 9 percent of the victims knew the thief was a family member or friend.

When people had their identities stolen, 43 percent contacted their credit card company, 26 percent reported the incident to local law enforcement agencies, and 22 percent reported the incident to a credit bureau; 38 percent did not report the incident to anyone. About half the victims reported the theft if it cost $5,000 or more. The average theft cost the victim $500 and about thirty hours to resolve—a total of $5 billion and 297 million hours in the year 2002.

What can we do? Well, there's the obvious. Don't give out your social security number. Carefully monitor your bank statements and credit card bills. Shred all documents with financial account numbers on them.

Beyond that? Not much. As one privacy consultant commented, "The problem is a little bit in the intractable category. For the most part, we rely on the good intentions of companies [that have customers' personal data]. But the empirical evidence says you cannot rely on that anymore. Bad things will happen. . . . Sooner or later it's going to happen. I don't know if there's really much we can do."[4]

The key to stopping this sort of identity theft, according to privacy advocates, is to stop trafficking of information among corporate affiliates, "because some companies have hundreds of businesses under the family umbrella. For example, a banking corporation might have a number of insurance, securities and real estate affiliates it does business with and financial data might be swapped among all." One advocate said, "If you cannot stop the traffic in your information among corporate affiliates, you don't have privacy in this nation."[5]

There's not much we can do about having our identity stolen.

I'll say.

But the real thieves aren't people like Cummings. When it comes to stealing one's identity, Cummings was a piker.

When someone from the corporate press or the FTC (or even you or I) talks about identity theft, what are they (or we) really talking about? We're talking about someone gaining access to your bank accounts and your credit cards. We may even be talking about someone stealing your (electronic) money. But what are they (or we) literally saying? That someone has "stolen your identity."

What just happened here? They (or we) are now identifying us with our bank accounts, with our credit cards, *with our finances*.

Who are you? Are you the bills you've racked up? Are you the money you have in the bank (or, to be precise, the magnetic blips on hard drives that are translated as money in the bank)? Are you your credit cards?

Who are you *as a person*? What are the qualities that make you who you are? What makes you part of various groups? What makes you different from all others? What identifies you and your individuality, your personality, personhood, character? Who is the real you, *in the flesh*?

• • • •

How is our identity stolen, worn down, erased? Who steals it, wears it down, erases it? How much of our identity is lost—stolen—in year after year of school, sitting in rows, begging the second hand to move faster, listening (or rather not listening) to teachers drone on and on about this or that fact or rule that doesn't matter to our own lives, to our identities—to *who we are*—as we absorb the main lessons: do not talk out of line, do not ask difficult or troublesome questions, always give ourselves away to those with the power to send us to the principal's office, to take away our recess, to give us bad grades, to hold us back in school if we do not toe (and tow) their lines?

And how much of our identity is lost or stolen when later we give ourselves away to bosses, to those with money or their surrogates, to those with the power to give us this money with which we have come to identify, to those who can fire us, who can make us homeless if we do not toe (and tow) their lines?

How much of our identity do we lose with each advertisement we see, each doctored photo that makes us loathe our bodies, our faces, makes us wish we were someone else? How much is lost each time we are told another lie by the corporate press, when each time we know it is a lie but cannot bring ourselves to believe our own truths? How much of our identity is torn away each time we actively tolerate the intolerable (the breast milk of every woman on the planet, for example, is contaminated not only with dioxin but 350 other chemicals such as heavy metals and those in perfumes, suntan oil, and pesticides[6])?

How much of our identity is lost or stolen as we give ourselves away—and this is especially true for those of us who still care to oppose all of this increasing control over our lives—to those who can arrest and imprison us if we don't do what they tell us? American Indian writer Ward Churchill recently talked about how everything in the United States (which calls itself the most free country in the world) is strictly regulated, and he challenged the audience to name any action a person could take that was not regulated by some law. Someone in the audience shouted out, "Smiling." Ward responded by citing a case where a man had been charged and convicted of "disorderly conduct" on the testimony of a police officer who, when asked what constituted the charge, replied, "Well, he smiled a lot."

How much of our identity is lost or stolen (or maybe just slips away) when we do not resist those whose policies are destroying life on this planet, those who believe their providence—their God, their Science—is leading them toward some strange utopia of absolute knowledge, absolute surveillance, absolute control? How much of our identity do we lose when we fear we are being watched by those at the center of the Panopticon? How much of our identity is necessarily stolen by the Panopticon's mere existence?

Each and every one of us is both victim and thief whenever we identify not with our own bodies, not with our humanity, not with our animality (we so often forget that we are animals, that we are primates), not with the landbases that support us, but rather with the very system that exploits us, that is killing us. To lose one's identity is to say, I am not a human whose body is rotten with the wastes of the industrial system, and whose mind is similarly polluted. I am not a human who loves this person and does not love that person, who has gifts and desires and insecurities and strengths and weaknesses all my own, but rather I am my job. I am a writer. I am an engineer. I am a scientist. I am an American. ("What do you think about our invasion of Iraq? Are you worried about our troops?" I ask, to which you may reasonably respond, "What do you mean? I didn't invade Iraq. They're not *my* troops. I do not identify with the U.S. government.") More simply still, I am my financial transactions. I am bits and bytes on hard drives all over the world. I am both simplified and fragmented. I am someone who lives in the outer ring of the Panopticon, hoping and praying—to whom?—that I do not do anything to attract the attention of the all-seeing eye at the center, the all-seeing eye on the back of the dollar bill—the dollar bill with which I have come to identify, the dollar bill that will almost undoubtedly soon have RFID chips in it. I am someone who wants to get through the day—get through my life—without unduly attracting the attention of those who have the power to punish me. This is how it is in school. This is how it is on the job. This is how it is in the culture at large. This is how it is and must be in the outer ring of the Panopticon.

Everyone and everything is unique. No thing—no being—is identical to any other. Every ponderosa pine is different from every other. Every caterpillar, dragonfly, human being, gust of wind has its own unique identity, unique attributes, unique beingness. Not its own unique identifying number. Its own unique identity.

Numbers are predictable, therefore controllable. Unique and willful beings are neither. The transformation of living beings to numbers, or rather our own transformation from people who perceive others as willful beings with unique identities to subjects who perceive others as objects—whether through science, economics, or the issuing of identity numbers—is all about control.

Electronic blips on hard drives are so much easier to control than wild and uncontrollable human (or nonhuman) beings. How much easier we make it for those at the center of the Panopticon if we allow them to rob us of our complex and willful identities, of our unpredictability, if we agree to identify ourselves with those numbers, and not with our bodies, our loved ones, our communities, the land where we live. We have become accomplices in the theft, snitches in the prison.

The Machine

Technological structures are "revolutionizing" human response by forcing life to conform to the parameters of the machine. . . . Even the shape of the child's developing brain is said to be changing. . . . What can transform to the computer, what can be transmitted by the technology, will remain; what cannot will vanish. That which remains will also be transformed by its isolation from that which is eliminated, and we will be changed irrevocably in the process. As language is reshaped, language will reshape daily life. Certain modes of thinking will simply atrophy and disappear, like rare, specialized species of birds. Later generations will not miss what they never had; the domain of language and meaning will be the domain of the screen. History will be the history on the screens; any subtlety, any memory which does not fit will be undecipherable, incoherent.

David Watson

There are many cultures in which it would be literally unthinkable for people to do what we do, to be who (or what) we have become. To be someone who would toxify and irradiate the totality of our environment, of our bodies. To systematically dismantle the ecological infrastructure of our (and everyone else's) landbase. To systematically enslave or eradicate our human and nonhuman neighbors. To attempt to control or annihilate everything around us. To attempt to control or annihilate all mystery.

Last night I dreamt I stood on a sandy hill overlooking the

ocean. The waves were big, but I thought I was safe. Suddenly I looked behind me and saw that instead of it being solid ground the sand sloped quickly again to water. I was surrounded by the sea. I turned back around just in time to see a mammoth wave rushing toward me. I felt the full weight of the ocean fall on me, then carry me away.

I often cry at the horrors perpetrated by our culture. I cry when I read of the eradication of creature after creature after creature, from Carolina parakeets to Siberian tigers to black-tailed prairie dogs to coho salmon to pikas to mahogany to monkey puzzle trees. I cry when I read of sweatshops and other forms of economically inspired misery. I cry when I read of the routine child abuse and everyday rape that characterize our culture (at least one in four women in our culture are raped within their lifetimes, and another one in five fend off rape attempts; what's more, many women believe that these figures grotesquely underestimate reality[1]). I cry when I read the prediction of George Orwell in *1984,* "If you want a picture of the future, imagine a boot stamping on a human face—forever."[2] I would add "nonhuman" as well.

Specific atrocities most often elicit my tears. With Orwell, however, the tears come from the recognition that he's got our culture's proclivities and direction pegged perfectly. In the same passage he wrote, "Do you begin to see, then, what kind of world we are creating? It is the exact opposite of the stupid hedonistic Utopias that the old reformers imagined. A world of fear and treachery and torment, a world of trampling and being trampled upon, a world which will grow not less but *more* merciless as it refines itself. Progress in our world will be progress toward more pain. The old civilizations claimed that they were founded on love and justice. Ours is founded upon hatred. In our world there will be no emotions except fear, rage, triumph, and self-abasement. Everything else we will destroy—everything."[3]

Yesterday I cried reading philosophy, for reasons similar to reading Orwell, because these philosophers so accurately articulate the pathology of our culture. But this was even worse because they do so approvingly. No, I wasn't reading the *Wall Street Journal* or anything else from the corporate press, which would have been bad enough. I was reading the philosophy that justifies current attempts to meld computers and humans, using something called nanotechnology, which is the science and engineering of materials and machines so small that they are invisible to the naked eye.

Before we get to the philosophy, let's talk a little more about nanotech. We can call it molecular engineering, we can call it the marriage of life and computers, we can call it the creation of green machines, or we can call it self-replicating biotech alchemy. It's been predicted to cure cancer, to end poverty by supplying the world with an infinite supply of energy and self-assembling materials, and to keep your pants stain-free.[4] Nanoparticles are already being used in cosmetics and industrial coatings.

Some of the gadgets and activities promised by the promoters of nanotechnology include nanoscale braille, biological RAM chips for your computers, DNA data storage, asteroid terraforming, anticancer nanomachines, space beanstalks, and multifuncton molecular manipulators.[5]

We are all by now unfortunately familiar with biotechnology, the splicing of genes from one being into another. Nanotech goes far beyond biotech's manipulation of genes; it involves the manipulation of molecules and atoms in the strange universe of quantum physics where things and energy, life and matter, interact in unpredictable ways. As one nanotechnology patent lawyer puts it, "On the nanoscale level, the traditional laws of physics cease to exist. Many times, even our clients don't understand what is going on with their inventions—and these are the leading researchers in America."[6] Nanotechnology researchers take advantage of these strange interactions to make stronger, more durable materials. Perhaps more significantly, nanotech will make possible the fusing of the biological and mechanical worlds.

Nanotechnology works with small materials. Really small. A nanometer is a billionth of a meter. Ten hydrogen atoms side by side make a line a nanometer long. A DNA molecule is about two and a half nanometers wide. A red blood cell is one-twentieth the width of a human hair, but it's 5,000 nanometers in diameter. It gives me a headache to just think about it: the individual components of a silicon transistor are about 130 nanometers across, but Intel can fit 42 million of them on one of its Pentium 4 computer chips.[7]

So, that's all great! We can all have fast computers to play more complex games (while the real world burns) because the geniuses at Intel have figured out how to get all those transistors onto a single computer chip. I can surf the Web, process these words, listen to historic folk songs recorded on a compact disk, and still have memory and space for the next zippy feature they sell me. I love it.

But it's not (yet) a perfect world, because those nasty atomtech critics tell us that "nano-sized bits are so small that they can penetrate your skin, get into your lungs, and travel through your body unmolested by the immune system."[8] Novels have been written about self-replicating nanobots run amok.[9]

All of which brings us finally to the melding of the biological and mechanical. The ultimate goal of nanotechnology, according to some of its proponents, is to meld humans and computers so that humans can at long last (insert mad scientist laugh here) live forever. As such, it is motivated by a fear of death, and the consequent fear and hatred of the body, the natural. Here's how the betterhumans.com Web site puts it, in an article entitled "Immortality": "The attribute of immortality is ascribed to an immaterial soul by some religions and philosophies. Others hold that immortality involves transmigration through successive humans and animals, with eventual absorption into an infinite being. Until recently, those who rejected such notions were given little in their place. Many secular philosophies hold that death is inevitable, natural and desirable. The only immortality available under this worldview is through actions that give us a place in the hearts and minds of future generations. Immortality through fame, however, is quite unsatisfactory to those who seek to avoid annihilation. It doesn't offer the never-ending continuation of consciousness and personal identity that humans have historically sought [*sic*]. It is mortalist, deathist thinking that characterizes aging and dying as natural and good. It encourages passivity, defeatism and abdication of responsibility for personal health and the future.

"Fortunately, celebrity is no longer the only alternative to supernatural promises. Today, immortality—and the lesser goal of indefinite lifespan—seems entirely possible in light of scientific theories on individuality and consciousness, the pace and direction of technological development and the existence of proven strategies for extending lifespan."[10]

How will this happen? The same author answers: "The best way to think of it is that our 'soul' is to our bodies and brains what music is to a CD: an arrangement of information. The CD doesn't matter—it can be copied an indefinite number of times. The music stored on the CD is what matters. If this seems a bizarre way to look at yourself, consider this: your body repairs itself throughout your life by taking material from the environment and discharging it through waste. Eventually, few of the atoms in your body will be the same as those from your youth. But for some reason your identity is continuous. How could this be? One explanation is that you have an immaterial soul. The other, upon which secular immortality is based, is that just like music on a CD, it's the arrangement of matter that counts. Just as destroying a CD destroys the music it contains, so too does the destruction of your body and brain obliterate you, by obliterating the neuronal connections in your brain that create the unique pattern of your identity. As the music dies, so too does your essence. So what could we do to achieve immortality? The same thing we would do to preserve the music on a rare, one-of-a-kind CD: make backup copies on a better medium."[11]

We could parse out the extraordinary and unfounded assertions in this paragraph, and the author's overt hatred of the body—right now a cat is licking my thumb, and this experience, this relationship, does not take place inside my mind, but between my physical thumb and the cat's physical tongue, and in the physical and emotional and intellectual and spiritual space that surrounds the both of us. But let us just note that even when it comes to music and to CDs, many musicians refuse to record their music because they feel that putting it on a CD enslaves or kills it, and that the music would not be music without the context that created it. All notion of context and relatedness is missing from this "betterhuman" perspective. I am who I am because of the redwood branches moving slightly in the breeze, the cat sitting on my lap licking my thumb, the other cat sleeping in the top drawer of the desk, the dogs sleeping outside the door, the hummingbird feeding at nearby flowers. It is absurd and pathetic that members of our culture generally define *who we are* as egos in sacks of skin—that is, when we aren't defining ourselves as our bank accounts—as opposed to the webs of relationships we share and the processes of those webs as they become and unbecome and rebecome; it is even more absurd and pathetic to drop the physical body as well. This is the endpoint and marriage of our culture's extreme solipsism and its hatred of the (uncontrollable) body.

Our betterhuman advocate continues, "With human identity and personality, there are various suggested techniques and technologies for achieving this. One of those often suggested is mind uploading, a technique that involves the transfer of our information patterns to a sophisticated supercomputer. While the technology to upload a mind doesn't exist yet, research in artificial intelligence, nanotechnology, and cognitive science, as well as developments in computer hardware, are taking us in the right direction. Mind uploading is also a possible side-effect of improvements in human-computer interfaces, as direct links between the brain and computer hardware could lead to a gradual merging of biological and nonbiological components of the mind. At some point, enough information could exist in the nonbiological portion that destruction of the biological brain has no impact on personality or identity."[12]

Destruction of the biological brain. Such a nice clean way to say the killing of the animal.

One nanotech philosopher, in an article entitled "Living Forever," points out, "The union of human and machine is well on its way. Almost every part of the body can already be enhanced or replaced, even some of our brain functions. Subminiature drug delivery systems can now precisely target tumors or individual cells. Within two to three decades, our brains will have been 'reverse-engineered': nanobots will give us full-immersion virtual reality and direct brain connection with the Internet. Soon after, we will vastly expand our intellect as we merge our biological brains with non-biological intelligence."[13]

In this rubric, "We can only solve the problem [*sic*] of mortality by moving from our biological substrate."[14] One writer, Alexander Bolonkin, formerly of the U.S. Air Force and NASA, writes, "An immortal person made of chips and supersolid materials (the e-man, as it was called in my [Bolonkin's] articles) will have incredible advantages in comparison with common people. An e-man will need no food, no dwelling, no air, no sleep, no rest, no ecologically pure environment. [This latter is certain to come in handy, all things considered.] Such a being will be able to travel into space, or walk on the sea floor with no aqualungs. His mental abilities and capacities will increase millions times. It will be possible to move such a person at a huge distance at a light speed. The information of one person like that could be transported to another planet with a laser and then placed in another body. Such people will not be awkward robots made of steel. An artificial person will have an opportunity to choose his or her face, body, good skin. It will also be possible for them to reproduce themselves avoiding the periods of childhood, adolescence, as well as education. It will not be possible to destroy an artificial person with any kind of weapons, since it will be possible to copy the information of his mind and then keep it separately."[15]

He is of course describing the goals of the übersoldiers' masters.

Bolonkin believes that "such transition to immortality (E-creatures) will be possible in 10–20 years. At first it will cost several million dollars and will be affordable only to very wealthy people, important statesmen, and celebrities. But in another 10–20 years, i.e. in the years 2020–2035, the cost of HEC (human-equivalent chip), together with the E-body, and organs of reception and communication, will drop to a few thousand dollars, and immortality will become affordable to the majority of the population of the developed countries, and another 10–15 years later, it will be accessible to practically all inhabitants of the Earth. Especially when at first it will be possible to record on chips only the contents of the brain, and provide the body for its independent existence later."[16]

The consequence of all this, according to Bolonkin, is that "the number of E-creatures will be growing and the number of people diminishing, till it gets to the minimum necessary for the zoos and small reservations. In all likelihood, the feelings that E-creatures may have towards humans as their ancestors, will be fading away, in proportion to the growing gap between the mental capacity of humans and electronic creatures, till they become comparable to our own attitude towards apes or even bugs."[17]

I cannot speak for his attitude toward apes, except to say that it seems to mirror the culture's attitude toward all of those it puts in zoos, in reservations, in the outer ring of the Panopticon or in the furnaces at Auschwitz, in the slash piles of clearcuts or back in the ocean dead as by-catch. I only know that I am an ape, a primate, an animal. I am not an electronic creature, a manufactured product of industrial processes. I began in sex, two apes coming together, as apes and other animals have done since their beginning, and will continue to do until a biology-denying and technology-obsessed culture eradicates them.

Bolonkin wants to get rid of procreative sex (which should come as no surprise since he wants to get rid of our bodies). He writes, "Another thing is quite obvious, too—that biological propagation will be so expensive, time-consuming, and primitive, that it will go into oblivion."[18]

Note that he wants to get rid of only procreative sex (and of course embodied sex) but not virtual sex with virtual hot chicks. Here's how it will work: "You do not need to worry that living in an electronic form will be dull and boring. It is vice versa, actually. When the information will be recorded onto other carriers, all human emotions, feelings and so on will also be carried over and preserved. In addition to that, the copies of certain emotions, pleasures, fears and so forth will be possible to record separately. After that, those separately recorded emotions and feelings can be given or sold to other people. Other e-men will have an opportunity to enjoy sex with a beauty queen, to experience the enjoyment of a sports victory, to take pleasure of power and the like."

If I see a picture of a naked woman on the Internet, do I know her? Have I now actually had sex with her? If people read books I've written, do they know me? If I listen to a CD of Beethoven's Ninth, do I now know Beethoven, and do I know the person who plays first viola? I once again cannot speak for Bolonkin, who does not write about relationships—because, of course, relationships do not exist in the atomistic, mechanistic, pornographic, split-off, virtual world of science and the Panopticon—but I do not want to download someone else's experience of "enjoy[ing] sex with a beauty queen," or more broadly of life. I want to live my own actual life, and if I were to "enjoy sex with a beauty queen," I would, odd as this may seem, want to know her, to have some sort of nonelectronic relationship with her, want to hear and see and smell and taste and feel her for myself, in the flesh (not only in the mind), in a particular place at a particular time in particular circumstances, in the context of a particular real (and personal) interaction with a real woman. Imagine that!

Let us return to Bolonkin: "All modern art is based on artists' aspiration to transcend [*sic*] their emotions, to make other people feel, what characters feel. Those works of art, which make that happen best, are considered to be outstanding and great. Electronic people will get those emotions directly. To crown it all, it will be possible to intensify those emotions, as we intensify a singer's voice now. Electronic people will have a huge world of all kinds of pleasures; it will be possible to know, what a dictator or an animal feels."[19]

Strangely enough, I already have the pleasure of knowing what an animal feels, because I am one. It is sad that he or anyone would write of wanting to feel animal pleasures mediated by a machine. If he means the pleasure of feeling what nonhuman animals feel (presumably not those in factory farms, presumably not those whose habitat is being stripped away as you read this), I desire only to know what they desire to let me know. The same is true for any beauty queen—absent in his discussion of sex with the beauty queen was of course any notion of consent on her part, but this is to be expected, as no one in science or pornography or the Panopticon asks permission from the object of one's attentions.

It is undoubtedly significant that the other pleasure Bolonkin mentions is that of feeling like a dictator. As well as being fueled by a fear of death, so too, as we've described, the Panopticon is fueled by a drive for power. As Orwell stated, "Already we are breaking down the habits of thought which have survived from before the Revolution. We have cut the links between child and parent, and between man and man, and between man and woman. . . . But in the future there will be no wives and no friends. Children will be taken from their mothers at birth as one takes eggs from a hen. The sex instinct will be eradicated. Procreation will be an annual formality like the renewal of a ration card [or copying a CD]. . . . There will be no curiosity, no employment of the process of life. All competing pleasures will be destroyed. But always—do not forget this, Winston—always there will be the intoxication of power, constantly increasing, constantly growing subtler. Always at every moment there will be the thrill of victory, the sensation of trampling on an enemy who is helpless."[20]

Bolonkin continues, "A soul living in such a virtual world will have all pleasures imaginable. It will be like living in paradise, as all [sic] religions see it. Computer chips of our time possess the frequency of more than two billion hertz. However, a human brain reacts to a change of environment only in one-twentieth of a second. This means that one year of life on Earth is equal to 100 million years of a soul living in the virtual world (paradise). Living in the virtual world will not be distinguishable from the real life. It will have a lot more advantages: you will have an opportunity to choose a palace to live in, you will have everything that you might wish for [except, of course, a body, an embodied life on a livable planet free from constant control by those who run the computers and electrical and technological infrastructures]. Yet, living in hell also becomes real. There is a hope that the ability to keep souls alive will be achieved by highly-civilized countries first. In this case they will prohibit torturing sinners, as they prohibit torturing criminals nowadays. Furthermore, criminal investigations will be simplified a lot, judicial mistakes will be excluded. It will be possible to access a soul consciousness and see every little detail of this or that action."[21]

This is the Panopticon made fully manifest. A simple scan of your computer disk and those in power know everything about you.

The vision is apocalyptic, as are the visions of all linear progressive religions, and is the vision being made manifest in the ecological and social collapse we are currently experiencing in the real physical world. As Bolonkin states, "Sooner or later religious teachings about soul, heaven and hell will become real. However, all that will be created by man. The so-called end of the world will also have a chance to become real, though. The religious interpretation of this notion implies the end of existence for all biological people (moving all souls onto artificial carriers, either to heaven or to hell)."[22]

Guess who decides who goes where, and guess on what grounds these decisions would be based.

The real purpose of it all, and we're quoting Bolonkin at such length not because he makes any sense (not that he makes less sense than other proponents of the cult of the Panopticon), but because he articulates so guilelessly and clearly the direction of the culture: "The goal of the mankind's existence is to create the Supreme Mind and to keep this Mind forever, no matter what might happen in the universe. The biological mankind is only a small step on the way to the creation of the Supreme Mind. The nature found a very good way to create the Supreme Mind: it decided to create a weak and imperfect biological mind at first. It took the nature millions of years to do that. The twentieth century was a very remarkable period in the history of the humanity. There has been incredible progress achieved, like never before. The scientific and the technological level of the humanity became sufficient for the creation of the artificial intelligence. This will be the first level of the Supreme Mind, when the human mind will make a step towards immortality. At present moment we stand on the edge of this process. It is obvious that biological people will not be able to compete with e-men by the end of this period. Common people will not be able to learn the knowledge that electronic people will get. The new cyberworld will be the only way for a human mind to survive. Feeble and unstable biological elements in a mind carrier or in its bubble will reduce its abilities and capacities a lot. If a common person will be willing to become a cyberman, then this cyberman will be more willing to get rid of all biological elements in his system and become like everyone. For example, there are no people in our present society, who would agree to become a monkey again. [Of course there is no need to become monkeys again: we already are, or rather, apes.] The Supreme Mind will eventually reach immense power. It will be able to move all over the universe, to control and use its laws. It will become God, if the notion of God implies something that knows and does everything. In other words, Man will become God."[23]

The truth is that it won't come to this, despite the $700 million a year the federal government pumps into the National Nanotechnology Initiative (one-third of which goes to the Pentagon, seekers of the übersoldiers)—making it one of the largest recipients of federal research money along with the "war on cancer" and Star Wars (the militarization of space)—and despite the enthusiasm of scientists, politicians, and bureaucrats who speak less flamboyantly yet just as forcefully as Bolonkin in favor of the processes. Dr. Mihail Roco, for example, head of the National Nanotechnology Initiative, says that nanotechnology will bring us a "new renaissance in our understanding of nature, means for improving human performance, and a new industrial revolution in coming decades," and that it will "fundamentally transform science, technology, and society. In 10–20 years, a significant proportion of industrial production, healthcare practice, and environmental management will be changed by the new technology." He also states that it will entirely transform a global society: "The effect of nanotechnologies on the health, wealth, and standard of living for people in this century could be at least as significant as the combined influences of microelectronics, medical imaging, computer-aided engineering, and man-made polymers [plastics] developed in the last century."[24]

Of course one might wonder if this transformation could turn out to be something other than positive, especially, as philosopher of technology Lewis Mumford observed in 1930, "once one drops the comfortable Victorian notion that all change is progress and all progress is beneficial."[25]

When something (inevitably) goes haywire, nanotechnology could, even according to some of its strongest boosters, lead to covering the entire planet with self-replicating "gray goo," what the experts call "global ecophagy," the eating of the earth. As K. Eric Drexler, author of *Engines of Creation: The Coming Era of Nanotechnology* and one of the most vociferous supporters of the nanotech vision, puts it, "Assembler-based replicators could beat the most advanced modern organisms. 'Plants' with 'leaves' no more efficient than today's solar cells could out-compete real plants, crowding the biosphere with an inedible foliage. Tough, omnivorous 'bacteria' could out-compete real bacteria: they could spread like blowing pollen, replicate swiftly, and reduce the biosphere to dust in a matter of days. Dangerous replicators could easily be too tough, small, and rapidly spreading to stop—at least if we made no preparation. We have trouble enough containing viruses and fruit flies."[26]

Despite this risk, the author of these lines still supports nanotechnology. His response to the potential "gray goo" killing of the

Earth? "I think talk about dangers is premature. These technologies are years away, and they'll have vast human benefits. Talking about dangers today will just bring out opponents and slow progress."[27]

We don't have to conjure gray goo to make nanotech deadly: rats exposed to 20-nanometer particles of polytetrafluoroethylene all died within four hours, while those exposed to larger particles of the same chemical all survived.[28] This shouldn't be surprising: Researchers have long known the dangers of nanoparticles—although they don't call them that, instead calling them "fines" and "ultrafines." As long ago as 1991, scientists at the Environmental Protection Agency estimated that fine particles kill sixty thousand Americans per year. Ultrafines are estimated to be ten to fifty times more dangerous, causing lung and cardiovascular disease, and probably promoting Alzheimer's and other forms of brain deterioration.[29]

You and I both know that these dangers won't stop those in power from pursuing this course. The proximate danger of the deaths of tens of thousands of Americans per year (and uncounted non-Americans, and uncounted nonhumans, both of whom matter even less than American humans to those in power) won't stop them, nor will the ultimate danger of gray goo eating the earth. Peter Montague, in his vital newsletter *Rachel's Environment and Health News,* outlines the "five-step pattern in this recent history of government-subsidized technologies. (1) It begins with a corporate decision to commandeer taxpayer funds to support the development of a new technology, after which government provides a long stream of subsidies, some in plain sight and many others hidden. (2) Next, we hear government (and corporate) hype about the limitless possibilities for increasing productivity, vastly improving the quality of life for everyone, ending poverty, curing cancer and so on. (3) Government then refuses to apply (or enforce) even the most common-sense regulations. (4) Government (in concert with the corporate sector) suppresses unwelcome information and ignores (or discredits) dissenting voices warning of trouble ahead. (5) Finally, government donates publicly-created knowledge and investment to corporate elites who then make profits galore for a decade or two until damage reports accumulate, the public catches on, and controversy engulfs the technology. The role of government throughout this phase is to act like a sponge and absorb blows from an angry public, suppress unwelcome information, discredit detractors, deflect demands for strict regulation, continue to hype the technology, simultaneously spending additional tens of billions of taxpayer dollars on elaborate (and contradictory) programs of blame, denial, cleanup, restitution, and defense against lawsuits."[30]

As Montague notes, "Nanotech has already entered stages 1 through 4 and is rapidly approaching stage 5."[31]

But nanotechnology and the marriage of computers and life will not fulfill their potentials for increases in industrial production and therefore decreases in the capacity of the planet to support life. This marriage will fail to take place not because governments suddenly decide not to kill their own citizens, nor because activists suddenly become effective. Instead, the natural world will stop these possibilities. With any luck for the rest of the planet, civilization will crash long before humans are stored on hard drives or CDs.

What must be grasped about all of this if we are to have any significant chance of altering the course (or rather helping the natural world to alter the course) of our deathly culture is that Bolonkin's vision has already come to pass. We are already living in the midst of it. Just as it would be a mistake to consider the Panopticon to be only a building of stone and glass and light and dark, it is a mistake to consider machines to be only artifacts made of iron and steel, and computers to be only metal boxes housing silicon chips. They are all much more. The Panopticon is a social arrangement, a way of life, a way of being in the world and relating to the world and to each other. The Machine, too, is a social arrangement, a way of life, a way of being in the world and relating to the world and to each other. And the Computer also is a social arrangement, a way of life, a way of being in the world and relating to the world and to each other. We are inside of the Panopticon, we are inside of the Machine, and we are inside of the Computer.

What does science do? It calls for everything to be measured. It calls for everything that cannot be measured to be ignored or destroyed, and everything that can be measured to be analyzed (according to the rules of science). It calls for calculations to be made as to how everything that can be measured and analyzed can best be used. It calls for those doing the measurements, calculations, and analyses (and most especially their masters) to rule over everything that can be measured. We are describing the methods and effects of science, not the conscious motivations of every scientist.

What is science for? To analyze. Why? To predict. Why? To reduce risk for those doing the calculations (and their masters) and to control those about whom (or, to use their lingo, which) these predictions are made. Why do they do this? So those performing these analyses and predictions can rule over everything they can analyze (and destroy everything they cannot).

Under this rubric, what is power? It is the ability to control outcomes.

What, then, is a bureaucracy? It is administration by rules, efficiency, and quantification. It is the administration of control.

What, then, is a culture administered by a bureaucracy?

It is a machine.

What are the necessary preconditions for the conversion of a living human community into a machine? Members of this community must begin to perceive themselves not as fluid threads in a complex and ever-changing web of relationships—where they may play this or that role as appropriate, necessary, and desired (by them and by others)—but as gears within cogs within gears in what they now perceive as a giant machine over which they have no fundamental agency, no loving stake. They must perceive their value no longer as inherent but as strictly utilitarian: they must be converted from human beings to workers. They must be made to perceive relationships as strictly hierarchical, where those closer to the outside of the Panopticon serve those closer to the inside, where rewards run from outside to inside, and only if there are any rewards left do some trickle back out. Everything must be perceived in terms of its short-term utility. Nothing must be given back.

Why would nanotech plants outcompete and overwhelm real live plants? Why would nanotech bacteria outcompete and overwhelm real live bacteria? Why does our machine culture outcompete and overwhelm real live cultures? Because machines are more efficient than living beings. Why are machines more efficient than living beings? Because machines do not give back. All living beings understand that they must give back to their surroundings as much as they take. If they do not, they will destroy their surroundings. By definition, machines—and people and cultures that have turned themselves into machines—do not give back. They use. And they use up. This gives them short-term advantages in power over the ability to determine outcomes. They outcompete. They overwhelm. They destroy.

Once people have been converted into cogs in their machine culture, the division of labor is increased, the skills of those in the outermost rings of the Panopticon are diminished, all heads are separated from hands (and hearts). Those in the innermost rings refuse to pay attention to anything that cannot be measured, and they convince everyone else to do so as well, if necessary at the point of a gun. They produce and convince everyone else to do so as well, once again at the point of a gun if those to be yoked to the machine have not been inculcated sufficiently to smile as they draw on their traces. Productivity is strictly defined in action (though it's best to not speak of this directly except when necessary) as the conversion of the living to the dead: living forests to two-by-fours; living rivers to hydroelectricity and from there to smelted aluminum (and from there to beer cans); living human beings to human resources. This conversion first takes place perceptually—subjects must stop perceiving others as subjects themselves and instead as objects—and then in the physical world. Efficiency is simply the rate and completeness with which the conversion takes place.

If people—or cogs that used to be people—are to be integrated into production, they must be recruited to be efficient. In practice, this means that nothing must stand in the way of production. Not leisure, not love, not a living landbase, not life on earth. That nothing human or animal is allowed to stand in the way of production can be surprising until we remember that production is, once again, the conversion of the living to the dead. That people are efficient simply means that life is not allowed to get in the way of its own killing.

Central to all of this is that it is more difficult by far to control diverse beings than it is to control objects that are all alike. Diversity must be destroyed. All cultures serving gods other than production—death—must be destroyed. All languages that do not serve this end must be forgotten. All creatures we can't use must be eliminated. All people must be standardized as well. (What do you think schooling is for?) One religion. One way of knowing the world. One economic system. One way of living on the land. If this language seems too strong to you, look around and ask what is happening to cultural diversity, to diversity of languages, to biodiversity, to all forms of diversity. They're disappearing. If you cannot perceive that, there is no hope for you. You will, however, always be welcome in the Panopticon.

I may as well be living in prison. For all practical purposes I already am. I spend most of my waking hours staring at a computer, moving in perhaps a five-by-five-foot space. Sometimes I walk to my mom's, and there I watch a baseball game. While there I generally confine myself to a similarly small space. Other times I get in a car, and I sit in a confined space while I drive to other buildings. I walk inside these buildings. I buy things. I come home. I sit in front of the computer. Later, I go to bed.

Sure, I walk outside in the forest, or I sit by the pond, or I watch the moon, but that's not *where I live*. I live inside this box. I live inside the world of the computer.

This morning as I sat outside for just a few moments before heading back inside to work at the computer, I heard a sound in the forest. A single branch cracked under heavy weight. Then silence. Then one squirrel began to speak. It was neither scolding nor yelling in fright. It was clearly gossiping. Another squirrel responded.

I do not know what creature walked through the forest and broke that single branch. The squirrels do, however. They know far more about this place I call home than I do.

Naturally they do. They live here.

I came back inside to work on the computer, to live inside my box.

I am alone in my cell. This would still be true if I were married. It would be true if I had a family. It would be true if I lived in a commune. This is because I would still be cut off from the world.

I am not alone in my predicament. Most of us in the industrialized parts of the world already spend more of our meager lives surrounded by machines, interacting with machines, supported by machines, dependent on machines, serving machines, than we do in meaningful communication and relationship with wild creatures. How many machines are within ten feet of you? How many wild plants or animals are within one hundred yards of you? Do you make yourself available to them? Do you know who they are individually? Do you have a feel for how well they may know you?

Step outside. Listen. How many machines do you hear? How many wild plants or animals?

Because our existential stance is so completely solipsistic— remember, the guard at the center of the Panopticon must always hold himself separate from those in the outer ring—it should come as no surprise that corporations are considered "persons" worthy of consideration, while flesh and blood nonhumans (and many humans) are not. Nor should it surprise us that the system better serves corporations—these legal machines for making money, amassing power—than it does mere "biological humans."

We are each and every one of us committed to, wedded to, enmeshed in the machine. The machine is grinding us up—each and every one of us—and will not stop until there is nothing left to grind.

Fear

"See," [the Pueblo man] Ochiway Biano said, "how cruel the whites look. Their lips are thin, their noses sharp, their faces furrowed and distorted by folds. Their eyes have a staring expression; they are always seeking something. What are they seeking? The whites always want something; they are always uneasy and restless. We do not know what they want. We do not understand them. We think that they are mad."

I [Carl Jung] asked him why he thought the whites were all mad.

"They say that they think with their heads," he replied.

"Why, of course. What do you think with?" I asked him in surprise.

"We think here," he said, indicating his heart.

Carl Jung

A chain of experience.

Fear of emptiness and discomfort lead to greed.

Greed leads to an obsession with getting what you want, which leads to putting acquisition and production above everything else. Frustrated greed leads to aggression, and the willingness to ignore others' feelings.

A byproduct of aggression is paranoia, because you fear that others are as aggressive as you are, which leads to an obsession with control (power over others) and security (protection of self). Of course the paranoia is often misplaced guilt over what you've done to others and to the community. (Just because you're paranoid doesn't mean people aren't upset. It's just that you have little understanding of how they feel.)

How do you maximize production? Through making everyone efficient, that is, through getting rid of barriers to production. One barrier to production is diversity. People, resources, machine parts must be interchangeable.

High technology is a tool of industrial production. Neither more nor less.

Bureaucracy is the administration of production and efficiency. It is nothing less than this, but it is more. Bureaucracy is the administrative means to eliminate feelings, ambivalence, and anything else that might interfere with production.

A lack of bureaucracy leads to a lack of efficiency.

A lack of efficiency leads to production not being maximized.

With production not maximized, the (neurotic) need to avoid discomfort through control is foiled. Fear returns.

What is the basis of most advertising? Fear, of course. Get a burglar alarm system. Cover up your bad breath, get the latest fashion, find a partner, then you need never again fear loneliness. Buy the latest piece of technology, and forget about your alienation.

Until it comes back. And then you can start over.

Not only must the natural world be standardized, and not only languages and cultures. Not only our lives are and must be standardized. Our thinking, especially, must be standardized. We must think the same things again and again and again. And again. We must remain in our cells, walking the same tight circles over and over. Our thoughts must never break out of their mechanical routine.[1]

Like on a CD—like in the nanotechnologists' wet dream—if someone pushes a button to start track one, we must always respond by playing the same song. By thinking the same song.

When another button is pushed, we must always play track two. Another button, another track. Volition must be excised.

Standardization leads to routine, leads to numbness. These are both necessary if we are to live in ways that are destructive of our own and the world's happiness.

Could someone else do your job? Are you replaceable? Are you encouraged to use the gifts that are unique to you in all the world? Are you encouraged to use them in the service of the land that gives you life?

What are we afraid of? We are afraid of our own impermanence, of our own death. We have been inculcated into this culture of the machine, where all rewards flow one way. We forget that our bodies, our very lives, are a covenant we enter into with the world, reminders of the reciprocal relationships that are who we are. I eat you now, you eat me later. All animals do this. All animals know this. Plants too, after a fashion. It is what everyone knows who participates in any real relationship.

Having denied this covenant, we hate and must destroy anything that reminds us of it, that reminds us of our own impermanence, our own death; that reminds us that nothing will make us ultimately happy because there is no ultimate satisfaction, only the daily pleasures (and pains) of living within the covenant (which themselves are wonderfully sufficient, bringing tremendous and fulfilling joy); that reminds us that we are not and can never be independent units in ultimate control of our own experiences or destiny. Our aggression inevitably leads us to the death we fear; not only is this aggression impotent to protect us from the inevitable, but these destructive patterns deliver precisely what they are supposed to protect us from.

Because we refuse to accept that there is no permanence, no ultimate satisfaction, and no independence, we fill our time with

Big Macs, with television programs, with the day's important busyness, and most crucially, with pretending to control. Wal-Mart exists because we are so afraid of facing ourselves without distraction. We are willing to throw away our lives and throw away the world by making and consuming plastic garbage. We attempt to distract ourselves from the emptiness of our existence—from the grayness of our prison cells and worldview—with brightly colored toys.

Lies are expensive to maintain. Our bodies know when we lie to them or to anyone else, so they must be deadened so they do not betray the nasty game we are playing. But our bodies still know, even when our emotions are numbed and our thoughts confused. We are not machines. We are animals who live and die and breathe and love and hate, who want to rest and to enjoy life and who do not wish to be cogs or slaves, who feel pity and do not wish to convert the living to the dead. And so we must be deadened, to make sure we do not slide back into our bodies. And then we must be watched, so that we do not forget that we are dead. Thus the omniscient God. Thus the Panopticon. Thus "the collective terror that identifies democracy with chaos and insecurity."[2]

In the modern world, not only must workers and slaves be monitored, lest they become inefficient—read *human,* read *animals*—and not only must malcontents (or those who may someday be malcontents) be monitored, lest they rebel, but so too must consumers be monitored, lest they cease to consume. And of course not only consumers must be monitored, but so too potential consumers. In an industrial system that depends on expanding markets, everyone is and must be a potential customer.

The answer? Monitor them all.
• • • •

The word *technology* comes from the Greek *technologia,* "systematic treatment." The Cambridge Advanced Learners Dictionary calls technology the study and knowledge of the practical, especially industrial, use of scientific discoveries.

Here is another definition. *Technology:* That which separates us from nature, that which leverages power.

Years ago I asked American Indian writer Vine Deloria, author of *God is Red; Custer Died for Your Sins;* and *Red Earth, White Lies,* about the difference between civilized and indigenous ways of being.

He said that the indigenous way is about seeing that "everything humans experience has value and instructs us in some aspect of life. Because everything is alive and making choices that determine the future, the world is constantly creating itself, and because every moment brings something new, we need to always try to not classify things too quickly. All the data must be considered, and we need to try to find how the ordinary and the extraordinary come together, as they must, in one coherent, comprehensive, mysterious story line. With the wisdom and time for reflection that old age brings, we may discover unsuspected relationships that make themselves manifest in our consciousness and so come to be understood.

"In this moral universe, all activities, events, and entities are related, and so it doesn't matter what kind of existence an entity enjoys—whether it is human or otter or star or rock—because the responsibility is always there for it to participate in the continuing creation of reality. Life is not a predatory jungle, 'red in tooth and claw,' as Westerners like to pretend, but is better understood as a symphony of mutual respect in which each player has a specific part to play. We must be in our proper place and we must play our role at the proper moment. So far as humans are concerned, because we came last, we are the 'younger brothers' of the other life-forms, and therefore have to learn everything from these other creatures. The real interest of old Indians would then be not to discover the abstract structure of physical reality, but rather to find the proper road down which, for the duration of a person's life, that person is supposed to walk."

That all made sense to me.

He continued, "I would also say that one of the major differences between Western and indigenous religions is that aboriginal groups have never had any need for a messiah. Not only is there no need, but in fact there really is no place for one in the cosmos."

"Why is that?"

"If the world is not a vale of tears, there's no need for salvation. Indians know nothing of a wholly different world—a heaven—compared to which this world is devoid of value. Indian religion is instead concerned with, as anthropologist Robert N. Bellah noted, 'the maintenance of personal, social, and cosmic harmony and with attaining specific goods—rain, harvest, children, health—as men have always been.'

"A commonality among all civilized religions is that of transcendence. But what I'm trying to get at is that in the North American Indian tradition individuals do not transcend themselves; they simply learn additional things about the single reality that confronts them.

"I think there are some very important questions we need to ask about religion in the dominant culture: Why do Western European peoples, and by extension before them the Near Eastern peoples, need a messiah? Why is their appraisal of the physical world a negative one? Why do their societies suffer such perennial and continuing crises? Why do they insist on believing that ultimate reality is contained in another, almost unimaginable realm beyond the senses and beyond the span of human life?

"I don't understand it. Religion as I have experienced it isn't the recitation of beliefs but a way of helping us to understand our lives. It must, I think, have an intimate connection with the world in which we live, and any religion that promotes other places—heaven and so on—in favor of what we have in the physical world is a delusion, a mere control device to allow us to be manipulated."

I asked, "What, then, in the Indian perspective, is the ultimate goal of life?"

He said, "Maturity. . . ."

"By which you mean . . ."

"The ability to reflect on the ordinary things of life and discover both their real meaning and the proper way to understand them when they appear in our lives.

"Now, I know this sounds as abstract as anything ever said by a Western scientist or philosopher, but within the context of Indian experience, it isn't abstract at all. Maturity in this context is a reflective situation that suggests a lifetime of experience, as a person travels from information to knowledge to wisdom. A person gathers information, and as it accumulates and achieves a sort of critical mass, patterns of interpretation and explanation begin to appear. This is where Western science aborts the process to derive its 'laws,' and assumes that the products of its own mind are inherent to the structure of the universe. But American Indians allow the process to continue, because premature analysis gives incomplete understanding. When we reach a very old age, or have the capacity to reflect and meditate on our experiences, or more often have the goal revealed to us in visions, we begin to understand how the intensity of experience, the particularity of individuality, and the rationality of the cycles of nature all relate to each other. That state is maturity, and it seems to produce wisdom."

We were both silent a moment.

He concluded, "Because Western society concentrates so heavily on information and theory, its product is youth, not maturity. The existence of thousands of plastic surgeons in America attests to the fact that we haven't crossed the emotional barriers that keep us from understanding and experiencing maturity."[3]

You may have heard the phrase, "Kill them all, and let God sort them out." Here is where it comes from.

The policy of the Catholic Church has been and continues to be *Nulla salus extra ecclesium,* which means "Outside the Church there is no salvation." This statement manifests not only the eradication of diversity that characterizes our culture but also inadvertently suggests the misery our culture leads to: outside this oppressive culture there is no need for salvation. Although this is still Church policy, we can modernize this statement by saying, "Outside Science there is no knowledge," or "Outside Technology there is no comfort," or "Outside Capitalism there are no economic transactions," or "Outside Industrial Civilization there is no humanity," or right to the point of this book, "Outside the Panopticon there is no security."

When those in power say that outside the Church (or Science, or Technology, or Capitalism, or Civilization, or the Panopticon) there can be no salvation, they are lying. What they are really saying is that you better not escape, because if *you* are outside of this Church, your continued existence (and happiness) will shake their own belief in the notion that the Church is good for them. Thus the one who can be saved only if you remain in the Church is the Church itself. If you leave, it ceases to be the arbiter of all meaning and the source of salvation. The salvation of the Church requires not only your belief and participation, but if you leave it requires your death, and beyond that, your annihilation. In industry, science, religion, and other institutions, diversity must be eliminated.

Thus the constant violence done by the believers in the Church (Science, Civilization, Capitalism, whatever name the sickness goes by at that moment) against all disbelievers, whether they are indigenous peoples, heretics, or nonhumans.

Early in the thirteenth century, Pope Innocent III gave "orders of fire and sword" against heretics all over Europe, collectively remembered today as Cathars. In just one French city, Beziers, one hundred thousand heretics known as Waldensians or Albigensians—believers in a widespread form of gnosticism—were murdered. After the sacking of the city, the inquisitors had a few hundred captives on their hands. Although the conquerors had promised, as they so often do, that those who surrendered would be allowed to live, once their power was in place the conquerors changed their minds, as once again they so often do. Many of their prisoners claimed to have never been heretics. But just in case they were lying, the inquisitors responded, "*Neca eos omnes. Deus suos agnoset,*" or "Kill them all. God will know His own." They hanged fifty, and burned the rest.[4]

Some scholars argue that the story is apocryphal because the phrase did not appear in any historical documents for another sixty years, and then only in a document written by someone "with an ardent imagination and very little concern for historical authenticity." As one skeptic writes, "No historian since 1866 has subscribed to the famous 'Kill them all;' but story writers still use it, and that is enough to prove how slow scientific acquisitions are in this regard to penetrate the public domain."[5] Actually, what this controversy proves to me is that there are a lot of people who will (feverishly) construct minor arguments to distract themselves and others from seeing the larger movement of our culture. Note that the scholarly controversy isn't over whether the Catholics massacred a hundred thousand people in Beziers—which they did—but whether anyone uttered this phrase.

I don't care so much about words as actions. For example, I care less about the pronouncements of clearcutters than I do about their actions: they kill all the trees and let economics sort them out, with fiscally valuable trees going to the mill and trash trees going into the slash pile to be burned. Industrial fishing?

They kill them all and do the same: fiscally valuable fish go in the freezer, and trash fish (birds, mammals) go overboard as bycatch. And humans? Count the number of nonthreatened indigenous cultures still living traditionally on their traditional homelands. If you get to one you've gone too far.

"Kill them all, and let God sort them out." We can alter the statement slightly and apply it to the Panopticon. The Panopticon is Technology's Church. Outside this Church there can be no salvation, and the word from on high is this: Monitor them all, let God sort them out.

God monitor us all.

Rationalization

> [This new type of man] turns his interest away from
> life, persons, nature, ideas—in short from everything
> that is alive; he transforms all life into things,
> including himself and the manifestations of his
> human faculties of reason, seeing, hearing, tasting,
> loving. . . . The world becomes a sum of lifeless arti-
> facts; from synthetic food to synthetic organs; the
> whole man becomes part of the total machinery that
> he controls and is simultaneously controlled by. He
> has no plan, no goal for life, except doing what the
> logic of technique determines him to do. He aspires to
> make robots as one of the greatest achievements of his
> technical mind, and some specialists assure us that the
> robot will hardly be distinguished from living men.
> This achievement will not seem so astonishing when
> man himself is hardly distinguishable from a robot.
>
> *Erich Fromm*

Scientists—and more broadly all believing members of this culture—often pride themselves on being rational and sometimes
sneer at indigenous (noncivilized) cultures as being not rational,
not reasonable, and based on superstition as opposed to "solid scientific observation," whatever that means. The truth, however, is
that this culture collectively and its members individually aren't
very smart. Let's be frank, we're pretty fucking stupid.

Let's examine some evidence. How rational is it to put poisons
on your own food? In the last seventy years worldwide annual
pesticide use has gone from zero to five hundred billion tons.
Cancer rates have risen, as have a host of other problems. For

example, children raised in areas with higher than "normal" pesticide use are more likely to suffer retarded physical and mental growth: they are made small, sickly, and stupid. How smart is it to poison your own children, observe that they are poisoned, and then keep doing it (or allowing it to be done)?

We already mentioned the prevalence of dioxin (and other carcinogens) in every mother's breast milk but did not yet mention that every stream in the United States is contaminated with carcinogens. At a talk I gave recently, I asked how many people in the audience had suffered the loss of someone they love because of cancer. All one hundred and fifty of us (including me) raised our hands. How rational is it for those in the Panopticon's inner circle to poison us and the ones we love, and themselves and the ones they love? How rational is it for the rest of us to not resist?

The stupidity continues: taxpayers pay to deforest the planet and pay to vacuum the oceans. Taxpayers, especially poor taxpayers, pay for police to put their sons and daughters in prison and pay for guards to keep them there. Prison guards get pay raises while the young are turned away from college.[1]

I've long had a habit of asking people if they like their jobs. About 90 percent say no. What does it mean that the vast majority of people spend the vast majority of their waking hours doing things they'd rather not do? Who would create a system so stupid? How stupid are we all to allow it to continue?

And it's simply not true to say that all cultures have toxified the air and water, deforested the land, murdered the oceans, or even that the members of all cultures have worked as hard as we do. Not only do indigenous cultures maintain livable landbases but they have far more leisure (in other words, killing the planet is damn hard work).

Rape is another thing that is really stupid (and of course rape is not "a product of our biology," as many apologists for our current rape culture would have it; many cultures have—or rather

had before they encountered ours—very low to essentially nonex-
istent levels of rape). Apologists for our rape culture often talk
about the importance of rape as a biological means of men
"spreading their genes," and of "a show of force" that lets women
know they are "better off bonding [*sic*] with a strong male." By
now I'm sure you can see what is missing: relationship. Even
from the perspective of "getting off" rape is really counterproduc-
tive: many of the traumatized women I've known have been far
less sexually free than they were prior to being violated. Even if a
man were to not care about the pain he inflicts, when nearly half
(or more) of women have been violated, there's an awful lot of
messed up sexuality in the air. That doesn't make for a particularly fun time for all concerned. The rapists may "get off"
through use of power over these women, but what's missing?
Let's say this all together: relationship. Sure, we can make the
argument that these levels of rape amount to a systematic program of terror, and that men derive the "benefits" of terrorized
women's subservience, but even the undeniable truth of this argument doesn't make the act of rape any less stupid, unless of course
you're more interested in having slaves than friends (which is,
you guessed it, the point of this wretched system). Similarly, child
abuse is stupid. What sense does it make to beat or rape one's
child? Over half a million American children are killed or
injured each year by their parents or guardians. Is that rational?

Well, the answer to that—and to all of the above questions—
depends on what you want. If you want servants, perhaps you can
perceive it in your best interests to terrorize others into serving
you. This is as true for terrorizing women as it is for children as it
is for the poor as it is for nonwhites as it is for nonhumans as it is
for any others to be enslaved. Of course this only works if you keep
blinders on to prevent you from perceiving not only their pain but
your own loneliness at having cut yourself off from relationship.

All of which finally brings us to a way in which our culture is extremely rational: there does happen to be one definition under which our culture is as rational as it pretends to be, which is that rationalization is the deliberate elimination of information unnecessary to achieving an immediate task.[2] If you want a culture full of terrorized slave-women on your hands, rape may be one way to accomplish that. If you want a culture full of terrorized slave-children on your hands, beatings may be one way to accomplish that. If you want a continent full of terrorized potential slaves on your hands, colonialism (currently administered with international "debt instruments" and "structural adjustment programs") may be one way to accomplish that. If you want a natural world full of terrorized slave-nonhumans on your hands, our entire industrial civilized way of life may be one way to accomplish that. You just have to not mind turning yourself into a machine (and, if you have difficulty ignoring the harmful consequences of your actions—that is, difficulty remaining a machine—you just have to remember to grab some of those "morning-after pills for regret" from the nearby übersoldiers).

To make this slightly more specific: If your goal is to maximize profits for a major corporation, all you need to do is ignore all considerations other than that. If your goal is to maximize gross national product (that is, the rate at which the world is converted into products), then all you need to do is ignore everything that might stand in the way of production.

As we see.

All of this reminds me once again of Bolonkin's vision. If the nerds have their way, we can all be turned into transhuman machines that have been permanently programmed to exclude information that isn't efficient, information like awareness of our own feelings, awareness of others' feelings, any broad perspectives, and awareness of and concern over the consequences of our actions.

Does this sound familiar?

There really is no need for us to be uploaded into machines, into computers. That's already what we are.

That's not entirely accurate. It's more accurate to say that the technologies we create mirror our own psyches. Only poisoned psyches could and would create and use five hundred billion tons of pesticides each year. Only emotion-dead psyches could and would create and use emotion-dead technologies. (And are we so enculturated, tradition-bound, and uncreative that we can't even conceptualize the possibility of technologies that aren't emotion-dead, excluding of course pathetic attempts to give computers emotions that will in the end be as phony as the faux sexual interest of women who've been terrorized into saying yes when they mean no? In contrast, many of the indigenous perceive drums, for example, and fires, as living, sentient beings with wants and desires all their own.) Only psyches driven by obsessions to control could and would create puppetrats they could control through their emotion-dead machines.

Jeremy Bentham had a nightmare of the perfect prison. Michel Foucault showed that the nightmare wasn't a building, but an administrative function that operated in schools, hospitals, and other institutions, indeed in the culture at large.

Early analysts of the industrial state, such as Karl Marx and Max Weber, described parts of this nightmare, and before we go too much further with our own exploration, it might be useful to survey their contributions.

More than a century and a half ago Marx and his coauthor Friedrich Engels described the evils and misery—the systematic dehumanization—that the industrial system causes. They saw that the bourgeoisie—that is, those whose primary concerns are commercial and industrial—have "converted the physician, the lawyer, the priest, the poet, the man of science, into its paid wage laborers. The bourgeoisie has torn away from the family its sentimental veil, and has reduced the family relation into a mere money relation."[3]

Marx and Engels recognized that within this system, workers are only another commodity, another resource: "Owing to the extensive use of machinery, and to the division of labor, the work of the proletarians has lost all individual character, and, consequently, all charm for the workman. He becomes an appendage of the machine. . . . Masses of laborers, crowded into the factory, are organized like soldiers. As privates of the industrial army, they are placed under the command of a perfect hierarchy of officers and sergeants. Not only are they slaves of the bourgeois class, and of the bourgeois state; they are daily and hourly enslaved by the machine, by the overlooker."

Marx and Engels described, not entirely disapprovingly (and at this remove in language that is shockingly ethnocentric), the inevitable expansion of the machine: "In place of the old wants, satisfied by the production of the country, we find new wants, requiring for their satisfaction the products of distant lands and climes. In place of the old local and national seclusion and self-sufficiency, we have intercourse in every direction, universal inter-dependence of nations. And as in material, so also in intellectual production. The intellectual creations of individual nations become common property. National one-sidedness and narrow-mindedness become more and more impossible, and from the numerous national and local literatures, there arises a world literature. The bourgeoisie, by the rapid improvement of all instruments of production, by the immensely facilitated means of communication, draws all, even the most barbarian, nations into civilization. The cheap prices of commodities are the heavy artillery with which it forces the barbarians' intensely obstinate hatred of foreigners to capitulate. It compels all nations, on pain of extinction, to adopt the bourgeois mode of production; it compels them to introduce what it calls civilization into their midst, i.e., to become bourgeois themselves. In one word, it creates a world after its own image."

Something ignored in this passage is that most often the "heavy artillery" of cheap commodities can only do its dirty work after honest-to-goodness heavy artillery has pounded a people into submission; after the target culture has been destroyed, most often at the point of a gun, and its members given the choice between Christianity (or Capitalism, or Science—*Nulla salus extra scientiam*—or Consumerism) and death; and after their landbase has been taken from them, its nonhuman members (and often its human members) converted to resources, and these resources stripped away.

Finally (for our analysis) Marx and Engels also saw that a fundamental imperative of our culture is the centralization of power, and they described the centralizing effects of industrialization: "The bourgeoisie keeps more and more doing away with the scattered state of the population, of the means of production, and of property. It has agglomerated population, centralized the means of production, and has concentrated property in a few hands. The necessary consequence of this was political centralization. Independent, or but loosely connected provinces, with separate interests, laws, governments, and systems of taxation, became lumped together into one nation, with one government, one code of laws, one national class interest, one frontier, and one customs tariff."

All of this is what we see around us: globalization of the industrial system, of the machine.

Much of what Marx and Engels called for in *The Communist Manifesto* has come to pass, with no need for their much-vaunted Proletarian Revolution. They called for the abolition of inheritance and the abolition of child labor in factories, which haven't happened, but also for a graduated income tax; a central bank;

centralized communication and transport; free public education; and the end of property in land. Now, before you private property patriots insist that the end of property in land has not taken place, and will take place only after the commies pry your second-amendment-guaranteed guns from your cold dead hands (not that I have a problem with gun ownership: as the bumper sticker says, "When automatic weapons are outlawed, only the FBI will have automatic weapons"), consider their next words, and realize not only that you're too late, that private property has already been done away with, not by godless commies but by capitalists who are overseen by providence, but more importantly that you've been snookered yet again by those in power: "You are horrified at our intending to do away with private property. But in your existing society, private property is already done away with for nine-tenths of the population; its existence for the few is solely due to its non-existence in the hands of those nine-tenths."

Marx and Engels anticipated much of what the industrial system would accomplish, and though they also described the soot and grime of the factories, they did not have real ecological understanding. This is not surprising, since almost nobody in our culture, even a century later, even environmentalists, has anything remotely approaching true ecological understanding. In fact we avoid it—rationalize it away—whenever possible. *The Communist Manifesto* called for factories owned by the state; bringing "waste lands" into cultivation; the establishment of industrial armies especially for agriculture; the combination of agriculture with manufacturing industries; the gradual abolition of the distinction between town and country through a more equable distribution of the populace over the country; and the combination of education with industrial production (most of which has also come to pass, damn it all). Marx and Engels believed that technology and "progress" were good, and that the proletariat would soon seize the means of production and abolish the class system so that everyone could enjoy the industrial fruits. To Marx and Engels, all relations were determined by the relations of production. All of history was only the story of class struggle, and once the workers had taken hold of the levers of the machines, class struggle (and history) would magically end. They did not understand that instead of the workers taking hold of the machines, the machines had already taken hold of the workers.

Max Weber understood that. He brought psychology and social dynamics into his understanding of the nightmare. He knew that society isn't moved only by material forces, and that human motivation isn't only rational or instrumental, but that motivation is a mix of rationality, values, emotion, and habit. He saw how modern society emphasizes the rational, the instrumental, the means to achieve certain goals, at the expense of everything else (including life itself). This narrowing of motivation in modern society has been fueled by the rise of bureaucracy and industrialization. Honing in, Weber saw that modern institutions, both public governments and private corporations, were becoming bureaucracies characterized by hierarchical authority, impersonality, rules, promotion by achievement, division of labor, and efficiency.[4]

Weber wanted to understand how traditional ways of living and being in communities had come to be abandoned in favor of the rational, the goal-oriented. He thought that nothing short of a religious motivation would be capable of overcoming people's natural tendency to work only until they were comfortable, rather than to pile up wealth for its own sake.[5] Capitalist propaganda aside, the allure of wealth has never been sufficient to make people work hard (especially when it's their hard work for someone else's wealth, which has always been how things happen in the "real world").

Here's how the religious motivation worked: The "waste of time," Weber wrote, became "the first and in principle the deadliest of sins."[6] He also wrote, "The religious valuation of restless, continuous, systematic work" became for this culture every human's perceived salvation, and the "most powerful conceivable lever for the expansion of . . . the spirit of capitalism."[7] "Outside the Church there is no salvation" became "Outside of Work there is no salvation."

At the risk of tipping our hand, we all know where this leads in the end, to *Arbeit Macht Frei* over the gates to hell.

Or to the factory. Sometimes you cannot tell the difference.

We need to not allow our analysis to become too sterile, too bureaucratic in itself. The pretense that the only things at work are psychology and social dynamics serves those in power. The Panopticon is ultimately based on force. It always has been and always will be.

Sure, when it's working well—when, to switch metaphors, the machine is well greased—the violence can safely hide in the background. That is the power of the Panopticon. Just yesterday I took my mom to Wal-Mart. Now, before you shout *hypocrite!* at my even visiting this temple of consumption, recognize that in this small town Wal-Mart has already wreaked its damage. My mom's telephone "died" the last time the electricity went out, and she "needed" another one. In this town her choices are Wal-Mart and Radio Shack. Would you like red or blue identity papers? My mom bought a new one, and it didn't work. I took her to exchange it. There was a line at the return counter, and it was a nice day, so I went outside to wait for her to finish. On one bench sat a woman eating a sandwich, and on another sat a man smoking a cigarette. I often prefer the company of bushes to humans anyway (especially the ones entranced by Wal-Mart) so I sat on the curb near some imprisoned pyracantha. Now here's the point: I could tell that those who walked by, especially Wal-Mart employees, were made uncomfortable with the fact that I was sitting in an unauthorized spot. And I know the problem was where I was sitting: I didn't have unauthorized long hair, nor unauthorized body odor, nor unauthorized dirty clothes, nor was I frowning in some unauthorized manner. But I could feel that people wanted me to move, and consequently I could feel myself wanting to move, to get back in line.

This is one way the machine works, when it is well greased. The same psychological pressures to conform would be at work were I instead standing with a pistol in my hand, pointing it at a Russian Jew kneeling beside a pit filled with writhing bodies, or with a chainsaw in my hand, pointing it at an ancient tree, or poised at a mass media magazine rack, choosing between *Soldier of Fortune, Penthouse,* or *Car and Driver.*

But, and here's the real point: what is the grease that smooths the machine? It is our own bodies and the bodies of others. Another way to get at this is by asking what happens to those who do not voluntarily enter the cells of the Panopticon, who do not voluntarily submit to the rules of the industrial-commercial-educational-security bureaucracy.

A hint: none of my students at Pelican Bay knocked on the doors begging to be let in. Well, precision requires that I amend that statement. They didn't knock the first time. Some few, having been what they term "institutionalized"—made incapable of surviving freedom—will commit crimes immediately on release to put themselves back in. I heard of one prisoner who escaped from minimum security a couple of weeks before his release date so he'd get sent back: prison had turned him into someone who could no longer survive freedom.

I'm sure you see how this applies on the larger, social scale.

But there still remains the question of what to do with those who will not become institutionalized. Of course we have the answers of what happened to the American Indians, and what continues to happen to the indigenous the world over: dispossession, dispersion, mass murder, genocide. And we also see what happens personally to those who resist: there are members of the Black Panthers who have been held in solitary confinement since the 1970s.

There's something missing amidst all this talk of those in power gaining information through darkened and lightened rooms in buildings that aren't actually buildings but instead metaphors for the whole culture. And that has to do with the actual gaining of information, where the rubber hits the road, as it were, or more precisely, where the rubber hits the flesh.

We need to be explicit about interrogation techniques employed by the CIA and associated groups. I have in front of me a CIA Torture Manual—oh, sorry, a Pain Compliance Manual, oh, sorry, this time a real title (and I'm not making this one up) *Human Resource Exploitation Training Manual*. I'm sure you can guess their contents. I also have in front of me the chapter from the 1963 CIA *KUBARK Counterintelligence Interrogation Manual* entitled "Coercive Counterintelligence Interrogation of Resistant Sources." These manuals are explicit: "The following are the principal coercive techniques of interrogation: arrest, detention, deprivation of sensory stimuli through solitary confinement or similar methods, threats and fear, debility, pain, heightened suggestibility and hypnosis, narcosis, and induced regression." No wonder they call it counterintelligence.

They go on to describe the advantages and disadvantages of each technique, and how each of them can be most effectively used to break their victims. The goal is to cause three important responses, "debility, dependency, and dread," that is, to cause their victims to "regress," that is, to lose their autonomy. As one manual puts it: "these techniques . . . are in essence methods of inducing regression of the personality to whatever earlier and weaker level is required for the dissolution of resistance and the inculcation of dependence. . . . As the interrogatee slips back from maturity toward a more infantile state, his learned or structured personality traits fall away in a reversed chronological order, so that the characteristics most recently acquired—which are also the characteristics drawn upon by the interrogatee in his own defense—are the first to go. . . . [R]egression is basically a loss of autonomy."

The point is to deconstruct the individual's self, or in short and in vernacular, to mindfuck victims until they give the perpetrators what they want. As the manual puts it: "Coercive procedures are designed not only to exploit the resistant source's internal conflicts and induce him to wrestle with himself but also to bring a superior outside force to bear upon the subject's resistance."

Every day, in subtle and not-so-subtle ways, we see these processes and purposes at work in the culture at large, whether from teachers, bosses, cops, politicians, or abusive parents who try to exploit our internal conflicts to increase their control, safe in the knowledge that if we refuse to be so exploited they will use force to achieve their same ends.

Wealth and consumption have come to form a never-ending circle: once work and wealth have been turned into a religion and made compulsive, the machine becomes self-propelling.

The dehumanizing impacts of bureaucracy also become self-propelling, as bureaucracy comes to dominate the quality of life and concentrate social, economic, and political power in the hands of a few. The concentration of wealth and power become inner and outer mirrors of the same dynamic: "This whole process of rationalization in the factory and elsewhere," Max Weber wrote, "and especially in the bureaucratic state machine, parallels the centralization of the material implements of organization in the hands of the master . . . [and it] takes over ever larger areas as the satisfaction of political and economic needs is increasingly rationalized."[8]

The problem, Weber understood, is nothing so straightforward as the existence of private property, or even a question of who controls the means of production. Weber and his colleague Robert Michels saw that the problem isn't—and I hate to break this news to all you old commies out there—capitalism, with its basis in private property and profit. Michels, a socialist himself, described how socialist organizations are also dominated by a few leaders, declaring his Iron Law of Oligarchy: "Who says organization, says oligarchy."[9]

In other words, regardless of mission statements that appear to be leaning left, right, up, or down, or whether they appear to be operating in the economic, political, or cultural domains, large organizations mean bureaucracy, and bureaucracy means hierarchy. Industrial society is too complicated for democratic governance. Once you accept the premises of our machine culture, centralization is inevitable, and efficiency must allocate the resources and rules control the machinery of the bureaucracy. This isn't a (merely undesirable) by-product of industrial organization: it's the *purpose. Organization:* from the Greek *organum,* tool or instrument.

The problem, then, as Weber saw, is that rationalization, order, and alienation are inherent characteristics of bureaucracy, and common to all forms of industrialization, socialist as well as capitalist: "The apparatus (bureaucracy), with its peculiar impersonal character . . . is easily made to work for anybody who knows how to gain control over it. A rationally ordered system of officials continues to function smoothly after the enemy has occupied the area: he merely needs to change the top officials."[10]

Every conqueror knows this. Don't destroy the bureaucracies. Use them.

That is, if you want to keep the machine running.

It doesn't matter who runs the machine, or even for what purpose: "From a purely technical point of view, a bureaucracy is capable of attaining the highest degree of efficiency, and is in this sense formally the most rational known means of exercising authority over human beings. It is superior to any other form in precision, in stability, in the stringency of its discipline, and in its reliability. It thus makes possible a particularly high degree of calculability of results for the heads of the organization and for those acting in relation to it. It is finally superior both in intensive efficiency and in the scope of its operations and is formally capable of application to all kinds of administrative tasks."[11]

Do you want to put a human on the moon? Assemble a bureaucracy. How about eradicating Jews, Slavs, Roma, and other *untermenschen*? Assemble a bureaucracy. Want to try to buy up land to protect it from being destroyed by industrial civilization? Assemble a bureaucracy. Want to try to dismantle the Panopticon? Assemble a bureaucracy.

But it's not quite that simple. Bureaucracies—like other machines—are better at some things than others: just as guns can't give birth and pesticides can't make plants, a bureaucracy cannot foster a vibrant community embedded in a thriving landbase. Unfortunately for everyone and everything on earth, machines—including bureaucratic machines—are better at destroying than nurturing, better at destroying than letting alone.

Weber also saw the irrationality of rationalization—that it works against values, emotions, and happiness. He wrote, "No machinery in the world functions so precisely as this apparatus of men and, moreover, so cheaply. . . . Rational calculation . . . reduces every worker to a cog in this bureaucratic machine and, seeing himself in this light, he will merely ask how to transform himself into a somewhat bigger cog."[12]

We too soon forget that we are not machines, that we are meant for something better than this. We search for rewards only within the system, having forgotten that there is a whole world waiting for us to remember that we are human beings and to drop out of—and destroy—the machine, and to rejoin the living world.

Weber continued, "It is horrible to think that the world could one day be filled with nothing but those little cogs, little men clinging to little jobs and striving toward bigger ones. . . . This passion for bureaucracy . . . is enough to drive one to despair. It is as if in politics . . . we were deliberately to become men who need 'order' and nothing but order, become nervous and cowardly if for one moment this order wavers, and helpless if they are torn away from their total incorporation in it. That the world should know no men but these: it is in such an evolution that we are already caught up, and the great question is, therefore, not how we can promote and hasten it, but what can we oppose to this machinery in order to keep a portion of mankind free from this parceling-out of the soul, from this supreme mastery of the bureaucratic way of life."[13]

Weber held little hope that we would be able to oppose the inexorable grinding of the machine. He thought industrial bureaucracy was so efficient, so powerful, that it was inescapable. He wrote, "The needs of mass administration make it today completely indispensable. The choice is only between bureaucracy and dilettantism in the field of administration."[14] Why is that? "The decisive reason for the advance of bureaucratic organization has always been its purely technical superiority over any other kind of organization. The fully developed bureaucratic mechanism compares with other organizations exactly as does the machine with the nonmechanical modes of organization." The result is that, as Weber states, "Without this form of (social) technology the industrialized countries could not have reached the heights of extravagance and wealth that they currently enjoy. All indications are that they will continue to grow in size and scope."[15]

Bureaucracies, it seems, are the gray goo that is eating the planet. Weber states, "Precision, speed, unambiguity, knowledge of the files, continuity, discretion, unity, strict subordination, reduction of friction and of material and personal costs—these are raised to the optimum point in the strictly bureaucratic organization."[16]

This is all true whether your bureaucracy is killing Jews, trees, or rivers. It is true whether you're running a professional army or a professional baseball team.

And the future? According to Weber, "Not summer's bloom lies ahead of us, but rather a polar night of icy darkness and hardness, no matter which group may triumph externally now." He also wrote, "It is apparent that today we are proceeding towards an evolution which resembles (the ancient kingdom of Egypt) in every detail, except that it is built on other foundations, on technically more perfect, more rationalized, and therefore much more mechanized foundations. The problem which besets us now is not: how can this evolution be changed?—for that is impossible, but: what will come of it."[17]

Indeed, what will come of us, and what will come of the living planet?

"No one knows who will live in this cage in the future, or whether at the end of this tremendous development entirely new prophets will arise, or there will be a great rebirth of old ideas and ideals or, if neither, mechanized petrification embellished with a sort of convulsive self-importance. For of the last stage of this cultural development, it might well be truly said: 'Specialists without spirit, sensualists without heart; this nullity imagines that it has obtained a level of civilization never before achieved.'"[18]

In the end, Max Weber was hopeless. He saw the "iron cage" that our culture has become and found it, like Orwell in his *1984,* inescapable.

For some, in fact for most, the iron cage is more like an Iron Maiden.

In the United States, one animal is killed in a vivisection lab every second. In Japan it's one every other second. In Europe, one every three seconds. The scientists who torture and kill these creatures for the most part probably do not consciously hate animals, hate bodies, hate the natural. Yet they force-feed agrochemicals and drano to dogs through tubes directly into their stomachs, and transplant the hearts and kidneys of pigs into the necks of baboons. They immobilize monkeys, lizards, cats, dogs, take off the tops of their heads. They break the necks of baboons. They addict macaques to cocaine, electroshock them if they will not use. They create superviruses that kill everyone they contact. They cut out portions of the brains of marmosets and leave these creatures as stupid as the experimenters themselves. They cut off the heads of live animals using scissors, then study their brains. They put live animals in freezers and let them try to claw their way out. They teach chimps American Sign Language, then put them in cages the size of cupboards: when the monkeys sign they want out, the scientists ignore their pleas, inject them with pesticides. They separate monkeys from their mothers, give them HIV, then put painful coils in their eyes to track where they look. Why? Ostensibly, at least, for knowledge, to understand how animals work, what makes them tick. Why? Ultimately to further control, and to further production.

It is not only the scientists who presumably do not consciously hate animals, but also the managers who run their departments, the vice presidents and presidents and CEOs who run the corporations, the janitors who clean the floors and windows, the cooks who staff the cafeterias, the electricians who keep the lights on, the accountants who count the beans, the police who provide security for the buildings. Still they all contribute to the torture of unimaginable numbers of animals. No, not to the torture of *numbers,* but to the torture of individuals, who live in cages, are tortured, and then are killed, or "sacrificed," to use the language of the scientific priesthood. They may be numbers to those inside the machine, but to those most intimately involved—to those who live and die in agony—there are no numbers, only lives. And that is the point of the machine.

Worldwide, approximately 214,000 acres of forest per day are destroyed. Most countries have already lost the vast majority of their natural forests, and rates of deforestation continue to rise, altered only briefly by the cycles of the global industrial economy. The people who cut down these trees do not for the most part at least consciously hate trees, hate the wild, and we can say the same for their bosses, and their bosses' bosses, all the way up the line. Yet the forests continue to be killed.

Ninety percent of the large fish are gone from the oceans, with no realistic hope of respite for the survivors. Those who run the factory trawlers presumably do not hate the oceans and those who live in them (Captain Ahab aside), nor do their bosses, their bosses' bosses, and so on.

Last summer, more than nineteen thousand people died in Europe from the hottest year on record, yet the articles about these deaths didn't even mention global warming. Do the corporate journalists hate truth, and do their editors, publishers, and so on? We think not (with a few notable exceptions).

Men and women in bureaucracies around the world do not torture. They do not deforest. They do not murder the oceans. They do not cause or ignore global warming. We merely push papers, attend meetings, do our jobs as well as we can, and then go home and try to relax (with television or chemical assistance if necessary).

The same was true in another holocaust. As preeminent historian of the Holocaust Raul Hilberg commented, "It must be kept in mind that most of the participants [of genocide] did not fire rifles at Jewish children or pour gas into gas chambers. . . . Most bureaucrats composed memoranda, drew up blueprints, talked on the telephone, and participated in conferences. They could destroy a whole people by sitting at their desk."[19]

And we have this account of other technicians in the Holocaust, "Specialists whose expertise normally had nothing to do with mass murder suddenly found themselves a minor cog in the machinery of destruction. Occupied with procuring, dispatching, maintaining, and repairing motor vehicles, their expertise and facilities were suddenly pressed into the service of mass murder when they were charged with producing gas vans. . . . What disturbed them was the criticism and complaints about faults in their product. The shortcomings of the gas vans were a negative reflection on their workmanship that had to be remedied. Kept fully abreast of the problems arising in the field, they strove for ingenious technical adjustments to make their product more efficient and acceptable to its operators. . . . Their greatest concern seemed to be that they might be deemed inadequate to the assigned task."[20]

The same is true now, except people are now destroying not just a whole people but a whole world by sitting at their desks, and by striving for ingenious technical adjustments that make their jobs more efficient.

Social critic C. Wright Mills commented on the strangeness of it all: "It is not the number of victims or the degree of cruelty that is distinctive; it is the fact that the acts committed and the acts that nobody protests are split from the consciousness of men in an uncanny, even a schizophrenic manner. The atrocities of our time are done by men as 'functions' of social machinery—men possessed by an abstracted view that hides from them the human beings [and nonhumans] who are their victims and, as well, their own humanity. They are inhuman acts because they are impersonal. They are not sadistic but merely businesslike; they are not aggressive but merely efficient; they are not emotional at all but technically clean-cut."[21]

The interrogation manuals often describe the techniques of breaking people with an absolute lack of attention to morality and humanity (and of course the same can be said of many manuals for teachers, bosses, cops, politicians, and parents), as though they're talking not about the destruction of human psyches (and bodies), but about how best to get to the grocery store: "Drugs are no more the answer to the interrogator's prayer than the polygraph, hypnosis, or other aids." Or this: Techniques are designed "to confound the expectations and conditioned reactions of the interrogatee," and "not only to obliterate the familiar but to replace it with the weird." When victims have been hammered with "double-talk questions" and "illogical" statements long enough, all sensible points of reference begin to blur, and "as the process continues, day after day if necessary, the subject begins to try to make sense of the situation, which becomes mentally intolerable. Now he is likely to make significant admissions, or even to pour out his whole story, just to stop the flow of babble which assails him."[22] Think about this role of babble the next time you wade through a daily newspaper, watch the evening news, drive down a street lined with billboards, or stroll through the mall.

Lewis Mumford, like Max Weber, saw the limitations of Marx's analysis and pointed out how Marx too-optimistically expected workers to rise up, and, more optimistically yet, hoped that if the workers somehow seized the levers of production this would actually accomplish something good. Marx, in Mumford's words, "was certain that the workers would be victorious; he did not admit that they might be more completely enslaved."[23]

Mumford held no such illusions. Part of his contribution to describing the nightmare that is industrial civilization was that he explored in far greater depth than Marx or Weber the premises of the culture's urge to destroy. For example, "The chief premise common to both technology and science is the notion that there are no desirable limits to the increase of knowledge, of material goods, of environmental control; that quantitative productivity is an end in itself, and that every means should be used to further expansion."[24]

His response was to insist that "what one must challenge is the value of a system so detached from other human needs and human purposes that the process goes on automatically without any visible goal except that of keeping the corporate apparatus itself in a state of power-making, power-yielding productivity."[25]

Mumford didn't point out the "tendency of mechanization and automation to form a self-enclosed system" in order to highlight some minor glitches, which are to be expected in any machine. "The point," he wrote, "is that the most massive defects of automation are those that arise, not from its failures, but from its indisputable [sic] triumphs. . . ."[26]

Mumford asked the same question that so many of us ask, which is, Why on earth would a culture do so many crazy, stupid, destructive things? His answer cuts through the cornucopian garbage we're all handed (and garbage it is: of what use are cool computer games on a planet being murdered?): "The desired reward of this magic [of automation] is not just abundance but absolute control."[27]

But you knew that already. Only "innocents," to use Mumford's overgenerous word (I would say "psychopaths," or maybe "the living dead"), could consider a "completely automated world society" to be the culmination of human evolution (all of evolution took place so we can watch television?). Instead, those of us with any sense at all recognize that the sort of automated future that civilization promises (or rather threatens) "would be a final solution to the problems of mankind, only in the sense that Hitler's extermination program was a final solution for the 'Jewish problem.'"[28]

We all know this in our bodies. Some of us even know it in our heads. Mumford commented in 1970 that "The notion that mechanical and scientific progress guaranteed parallel human benefits was already dubious by 1851, the year of the Crystal Palace Exhibition, and now has become completely untenable."[29] And yet in 1933 the title of the World's Fair in Chicago was *The Century of Progress*. The slogan over the gate? "Science explores, technology executes [certainly in more ways than one], man conforms."[30] This was during the Great Depression, that colossal glitch (but not failure) of the machine's economic system. In 1933 the unemployment rate was peaking at 25 percent (something's dramatically wrong with a system where three out of four people are working their lives away, yet the economic system still falters) and farmers were dumping milk while people starved. Technology was executing, but some men weren't conforming.

Mumford wasn't wildly optimistic about the future. He knew—as we all do—that there was no hope in proceeding "on the terms imposed by technocratic society . . . [with] its plans for accelerated technological progress, even though man's vital organs [and the rest of the world] will all be cannibalized in order to prolong the megamachine's meaningless existence."[31]

He knew also: "The ideology that underlies and unites the ancient and modern megamachines is one that ignores the needs and purposes of life in order to fortify the power complex and extend its dominion. Both megamachines are oriented toward death; and the more they approach unified planetary control, the more inescapable does that result promise to become."[32]

He didn't think change would be easy, saying, rightly enough, that it might take "an all-out fatal shock treatment, close to catastrophe, to break the hold of civilized man's chronic psychosis. Even such a belated awakening would be a miracle."[33]

But at least Mumford knew that more automation and technology wasn't the cure for too much automation and technology, which is more than we can say for those currently in power. He knew that the way out was that the "most solid-seeming structures and institutions must collapse as soon as the formative ideas [actually, the perceptual distortions and emotional projections] that have brought them into existence begin to dissolve."

Mumford was, in the end, at least slightly optimistic (and when looking at this culture one learns to take cautious optimism wherever one can find it), declaring that "for those of us who have thrown off the myth of the machine, the next move is ours, for the gates of the technocratic prison will open automatically, despite their rusty hinges, as soon as we choose to walk out."[34]

The Panoptic Sort

It seems to me the sailor's card, and not the sun, is the center of the universe. I am positive that the great war was fought, not for democracy and justice, but for no other reason than that a cop, or an immigration officer, may have the legal right to ask you, and be well paid for asking you, to show him your sailor's card, or what have you. Before the war nobody asked you for a passport.

B. Traven

In the 1990s, Oscar Gandy took up where Foucault left off by describing the "panoptic sort" as "the complex technology that involves the collection, processing, and sharing of information about individuals and groups that is generated through their daily lives as citizens, employees, and consumers and is used to coordinate and control their access to the goods and services that define life in the modern capitalist economy."[1] In other words, it's how those in power manage information that allows them to manage you, both collectively and individually. The more information they have, the more effectively they can manage you.

The basis of the panoptic sort is the remote, invisible, automatic, and comprehensive sensing of personhood and the classification, evaluation, and sorting of individuals into groups for efficient training, rehabilitation, or elimination, based on their value to the economic and political elite who control the sorting.

The most highly valued are the rich and other rulers; they are given the primary fiscal benefits of the sorting system. Also high in the hierarchy are those trusted strategists who can make sense of the vast information apparatus. Below them are technicians who are privy to the data collected by the surveillance machines. Below that are the people of the middle class who enjoy enough benefits so that their sense of privilege outweighs their nagging feeling of never quite reaching the top. (From the point of view of those who run the system, the value of the middle class is to provide the bulk of the surplus value.) Below the middle class are working-class people, who run and maintain the machines that produce the consumer goods. They, too, enjoy enough benefits to keep them at work, to give them the illusion that they are living a good life, and to keep them from looking for a different way to live. And, as Henry Ford saw early on, it is essential in an industrial system to give at least some of the workers enough pay to buy at least some of what they build, or else the system's inevitable overproduction has no outlet. Toward the bottom of the value scale are those who are "of little substance who carry the sick, bury the dead, clean and do many vile and abject offices."[2] But even the unemployed and the homeless are of some value to the system. For example, they keep wages low by making the working class fearful of losing their jobs and by making sleeping under a bridge seem the only alternative to the treadmill of rent or mortgage. Below the value scale altogether are those who will not partake of the benefits of the system: the hunter-gatherers, the subsistence farmers who own their own land, the gypsies, the odd free spirit who will not settle for mortgage and salary. These are worse than useless to the system, because they provide the system's servants with alternative visions and lifestyles. Because the existence of these alternatives cannot be tolerated, lest the servants become restless, those who live these alternatives must be banished from the servants' view, or destroyed altogether.

This is where passports come in. Passports have been used for centuries for many purposes: to differentiate economic classes and religious and ethnic groups; to control serfs; to control travel across borders; to enforce military service; to collect taxes or other debts; to limit the labor pool; to monitor or immobilize the poor, the homeless, the unemployed, the criminals; to expel unwanted foreigners; to monitor the movement of citizens within the homeland; and to specify where the holder could and could not reside.[3] The Russian nobility controlled the movement of its serfs by registering them with passports. Two hundred years later the Soviet state used passports to enforce a collectivization more stringent, disempowering, and deadly than anything the old nobles could have dreamed up.[4] Early in the twentieth century, the British used passports to control the movement of Eastern European Jews. The French used passports to keep young men from escaping military conscription. Internal passports were a primary tool of South Africa's apartheid system. The Rwanda genocide of the 1990s used ethnic identity cards introduced by the colonial masters.[5]

In times of unrest, revolution, and war, the use of passports is ratcheted up. New classes of people are required to carry them: anyone from the enemy nation (they might be a spy!), anyone eligible for military service (mustn't let them leave the country!), anyone who is traveling (who are they going to meet?!). These tighter standards are often not relaxed once peace returns. After World War I, the U.S. government made permanent the use of passports to exclude "the undesirable, the enemy of law and order, the breeder of revolution, and the advocate of anarchy."[6]

In fact, why not have a national identity card for everyone? This shouldn't be a problem, if, as the saying goes, you "have nothing to hide, and you need hide nothing." The classic argument in favor of the continued ratcheting up of restrictions after World War I was articulated by the British trade unionist W. L. George. In his April 6, 1919 *Sunday Times* editorial, he wrote that "secrecy is hateful; it breeds evils, and I would that all houses might be made of glass." Take note of what he has done here: by shaming the reader, he attempts to skirt the central truth that secrecy and evil come not from houses so much as from the fortresses and armories of bureaucracy. If you buy into his argument—and too many people do—it's easy to ask, Why shouldn't the identity card and a national register be seen as "hallmarks of efficient, modern bureaucracy, and not an emergency measure"?[7]

Early passports contained the holder's name and hometown. But names were easily changed[8] (despite the laws of European states forbidding it), so the traveler's permanent features such as scars or tattoos were also recorded. Well into the nineteenth century, branding and ear-docking were common ways to identify criminals, gypsies, indigents, and the insane. Jeremy Bentham, architect of the Panopticon, recommended that *everyone* be tattooed, as a "new spring for morality, a new source of power for the laws, an almost infallible precaution against a multitude of offences."[9] Bentham was an incurable optimist (of a sort); the technology for such a thorough system was more than a century away, and even in Bentham's time, branding and other gross methods were beginning to be frowned upon. Administrators faced a difficult problem: they needed something that didn't hurt so much when administered, and that didn't reveal the coercive power of the sorting machine.

Luckily for them, anthropometry (the study of human body measurements, especially on a comparative basis), fingerprints, and photography were soon discovered. And used. Within ten years of the invention of photography, the Swiss and German police had taken hundreds of thousands of photographs of gypsies and other unemployed (useless) and homeless (mortgage-free) people.

By 1840 France decreed that all nomads, those without "domicile or fixed abode," itinerants, and "romanies," who were accused of poaching, theft, bestiality, and stealing children, should have their height, chest, armspan, head, and left foot measured, their eye color and fingerprints recorded, and their photographs taken. It wouldn't do much good to measure their growing children, but everyone down to the age of two years was to have all ten fingerprints recorded.[10] The "romanies" were accused of stealing children, but who in fact was taking control of other's children's identities?

The bureaucrats' problem quickly became the sorting and retrieval of a particular photo from the hundreds of thousands that had been collected. So many unemployed people roaming the land, and so few bureaucrats! With their disregard for the supposed benefits of supposedly gainful labor, the unemployed certainly weren't to be trained. How, then, were they to be followed and controlled?

Discrimination is easy. So is building a database. Retrieval and sense-making are the trick. It's a big world out there, full of people and politics and uncontrolled behavior (and, until we eliminate them, lions and tigers and bears, oh my). Sorting them all to pick out certain ones is the big challenge. Our culture has opted for the easy way out: bring in the machines.

That's what the Nazis did. They had a problem. How were they to keep track of millions of Jews and other undesirables? They turned to IBM and its Hollerith machines. "As early as 1934, various government bureaus began to compile card catalogs identifying political and racial enemies of the regime, such as Freemasons, Jews, Sinti and Roma (Gypsies), and 'genetically diseased' persons. [Actually, governments had been doing this for more than a century, but no mind.] The 1939 census became the basis for a national register of Jews. That year, German census forms for the first time included explicitly racial [*sic*] categories. Jews were identified not only by religious affiliation, but by race as well. Within three years, the completed national register of Jews and some Jewish Mischlinge ('mixed breeds') was to become one of the sources for Nazi deportation lists. Most of those deported perished in the Holocaust. During the 1930s and 1940s, Hollerith machines were the best data processing devices available. The Nazi regime employed thousands of people in 1933 to 1939 to record national census data onto Hollerith punch cards. The SS used the Hollerith machines during the war to monitor the large numbers of prisoners shipped in and out of concentration camps. The machines were manufactured by DEHOMAG—Deutsche Hollerith Maschinen Gesellschaft or German Hollerith Machine Company, a subsidiary of IBM since 1922."[11]

Let's bring this up to date. Like the French passes for gypsies, the border-crossing and green cards that the U.S. government issues to Mexican citizens display fingerprints and photos. But in this dangerous post-911 world—or, to be more precise, in this world where those in power have seized upon the excuse of 911 to increase their power to surveil and control—those in power are getting much more sophisticated.[12] By October 2004, all foreign visitors to the U.S. are supposed to have visas and passports that include computer chips with facial recognition data. Travelers' fingers and faces can then be scanned to compare with the digital data on their passports. By 2006, all new passports issued to U.S. citizens are also to contain the chips. Currently there are 55 million U.S. passports, with another 7 million issued every year.[13]

The term used for all this technology is *biometrics*. It's just what it sounds like: the measurement of living beings. Each of us is unique, and various economic and political authorities (using their handmaidens, science and technology) are finding many ways to measure various bits and pieces of us. Your face, fingerprints, irises and retinas, wrist vein patterns, DNA, handwriting, voice, and gait can all be recorded, measured, analyzed, and identified. (Note, of course, that none of this is *who you are:* just as "identity theft" reduces us to our bank account, biometrics reduces us to our fingerprints.) Technically, much of this is a piece of cake, and the only reason the feds don't already have your wrist vein patterns on file (presuming they don't) is because, as *The New York Times* puts it, "the adoption of biometric technologies has been held back for years by concerns about privacy and reliability, along with a lack of uniform standards. But . . . policies and standards have begun to catch up with the technologies."[14]

"Policies and standards have begun to catch up with the technologies." Doesn't this phrase sound familiar? Isn't this how it always goes? Many who have explored technology, surveillance, and bureaucracy—we're thinking here of Max Weber, Jacques Ellul, Christopher Dandeker, and others—have come to the same realization, that the advance of technology is relentless and that it is impossible to control surveillance with laws and regulations. Who—or what—is in charge here?

Don't worry, though, whisper the voices inside your head (at least you think they're inside your head, but you're not quite so sure anymore): the passport chips will be encrypted, and even though it involves face recognition software, the system will be "very user-friendly; it's unobtrusive." This was said by Denis Shagnon, an official at the International Civil Aviation Organization, the United Nations (UN) agency that set an international standard for biometric passports in May 2003. He also said, "What was required was a globally interoperable biometric—one biometric that could be used worldwide and can be read worldwide."[15] Under the new standards, governments will be able to add additional biometric technologies to passports, like fingerprints or iris scans.

As its name indicates, the passport was originally used to authorize travel by sea. Thanks to the UN's International Labour Organization, more than a million workers on ships around the world will soon have biometric identification cards that store digital fingerprints and photographs in a two-dimensional bar code.[16]

The passport, being the property of the state, can be withdrawn at any time, thus revoking the privileges of travel, work, residence . . . and identity. Jeremy Bentham keeps reminding us that the central question is not your question, "Who am I?", but the one directed at you, "Who are you, with whom I [the bureaucrat] have to deal?"[17]

The interrogation manual: "The manner and timing of arrest can contribute substantially to the interrogator's purposes. What we aim to do is to ensure that the manner of arrest achieves, if possible, surprise, and the maximum amount of mental discomfort in order to catch the suspect off balance and to deprive him of the initiative. One should therefore arrest him at a moment when he least expects it and when his mental and physical resistance is at its lowest. The ideal time at which to arrest a person is in the early hours of the morning because surprise is achieved then, and because a person's resistance physiologically as well as psychologically is at its lowest."[18]

We don't track gypsies anymore,[19] but we're still interested in the unemployed, the poor, the hungry. In 1996, the Connecticut legislature passed a law requiring that the two index fingerprints of every existing and new welfare recipient be scanned and databased. With a computer, not an inkpad. No muss, no fuss. It would be done "to prevent people from receiving welfare benefits under more than one name or from receiving benefits improperly from more than one town or state program."[20] The poor people are still poaching from the king's larder, and it's got to stop. The fact that no pretense is made of stopping the far greater thefts from the treasury by corporations "receiving benefits improperly" hints at the real purpose of this information-gathering.

In standard bureaucratic form, a survey was taken of the new program: 88.4 percent of the respondents felt they had not been inconvenienced by the finger imaging process; 85.1 percent did not object to the process; and 87.1 percent felt that the process would indeed help prevent people from cheating the welfare program.[21]

Perhaps the welfare recipients felt themselves in no position to protest. Certainly had they spoken out against the program, they would have been reminded that the receipt of welfare benefits is purely voluntary, and if they did not wish to give their fingerprints, no one would put a gun to their head and force them to apply for assistance.

The state knew the program might be controversial elsewhere, since "public perception and the association of fingerprinting with the criminal element was pervasive." Using society's obsession with technology to gloss over what was happening, the Department of Social Services (DSS) had "early on . . . determined that the project should be referred to as Digital Imaging with the focus on the digital technology rather than fingerprinting alone." DSS's "opponents" characterized the system as "invasive and dehumanizing." DSS's response: a press conference during which the Connecticut governor, followed by state legislator Jeanne Garvey ("The Mother of Digital Imaging"), the DSS commissioner and deputy commissioner, and several other state officials, were the first to be "imaged." DSS is confident that "the demonstration of the application of the technology was straightforward and the value apparent to all."[22]

Who's next? The answer: Eighty-four thousand adult recipients of public assistance from 169 towns. The state believes "the connection of the system to ID card issuance and its possible future use in the nationwide EBT infrastructure would position the project as a model for the future of human services benefit delivery systems."[23]

Connecticut is small potatoes. The British national DNA database has two million profiles and more than five million sets of fingerprints. Authorities are seeking new powers that would allow police to take DNA and fingerprint samples whenever a person is arrested, not just if that person is actually charged with a crime. A court ruling states that the DNA of people who have been cleared of any involvement in a crime can be retained. The benefit to society? If a suspect's DNA is put through the system, 38 percent of crimes are solved, compared with an overall figure of 24 percent.[24] The U.S. Combined DNA Index System, established by the FBI in 1990, enables federal, state, and local crime labs to exchange and compare DNA profiles electronically.[25]

Not to worry, whisper those voices in your head. If you don't do anything wrong, you have nothing to fear, right?

Right?

More from the CIA manual: "The effectiveness of a threat depends not only on what sort of person the interrogatee is and whether he believes that his questioner can and will carry the threat out but also on the interrogator's reasons for threatening. If the interrogator threatens because he is angry, the subject frequently senses the fear of failure underlying the anger and is strengthened in his own resolve to resist. Threats delivered coldly are more effective than those shouted in rage. It is especially important that a threat not be uttered in response to the interrogatee's own expressions of hostility. These, if ignored, can induce feelings of guilt, whereas retorts in kind relieve the subject's feelings. Another reason why threats induce compliance not evoked by the inflection of duress is that the threat grants the interrogatee time for compliance. It is not enough that a resistant source should be placed under the tension of fear; he must also discern an acceptable escape route."[26]

Okay, so you aren't a criminal, and you aren't on welfare in Connecticut. But the authorities can still keep track of you—if you have a car or a cell phone.

Ninety-nine percent of the U.S. population can dial 911 in an emergency. A third to half of the 911 calls are now made using cell phones. But in one test, 15 percent of the 911 calls made on cell phones didn't get through. And only 1 percent of the agencies in charge of responding to emergencies can identify the location of a 911 call made from a cell phone.

To better protect and serve us, in 1999 the U.S. Federal Communications Commission (FCC) ruled that 911 operators should be able to identify cell phone numbers just like they do land line numbers—and that new cell phones should include global positioning system (GPS) technology so that a phone call made on a cell phone could be pinpointed within one hundred yards. The rules were to be in effect by 2005 but the need to upgrade both cell phone design and 911 center technology has slowed implementation of the new rules.[27]

The National Emergency Number Association (NENA) would like to change that. Who is NENA? They've been around for twenty years, and have seven thousand members and forty-six chapters. Their purported mission is to "foster the technological advancement, availability, and implementation of a universal emergency telephone number system." The "protection of human life, the preservation of property and the maintenance of general community security are among NENA's stated objectives."[28] But we might wonder at their real objectives when we learn that NENA's partners in public forums include representatives from the Department of Homeland Security, U.S. Department of Justice, the U.S. Congress, and the FCC. The NENA Business Alliance includes Nortel, Lucent, SBC, ATX Technologies, Motorola, Tel Control, and other high-tech telecommunications corporations. Do you believe these corporate personhoods have your security interests at heart?[29]

No cell phone? Not a problem for those in power. Do you drive a car? The new black boxes in cars are a classic example of technological innovation, function creep, the obsession for security, the drive toward standardization, bureaucratic compartmentalization, and the impotence of privacy law. Twenty-five million cars in the United States have event data recorders in them. The original and ostensible purpose of the recorder was to "monitor sensors and decide whether to fire air bags." So far so good. But we'll go step by step through what happened next to illustrate a process that happens with technology after technology. First, technological innovation: a new feature allows some data to be stored, such as the speed of the vehicle just before a crash. Next, function creep: General Motors has been gathering statistics from the recorders since the 1990s—always with the owner's permission, they say, since consumer privacy is a "top concern" with GM. Then, more function creep: the Vetronix Corporation sells machines that give outsiders the ability to read the recorders. Insurance companies and police agencies are interested. Then, the drive toward standardization: the National Highway Traffic Safety Administration (NHTSA) is considering whether the auto industry (and the allied industries of insurance, police, highway construction, and so on) should standardize the equipment. Bureaucratic compartmentalization: NHTSA says the standards are years away, and that it is the courts that should decide who can use the data in the recorders. Impotence of privacy law: personal injury attorneys, as well as police, are pointing out that your driving affects us all, and anyone on the road can see how fast you're driving, so you can't expect any privacy here. They say, "the privacy claim is just an excuse for keeping people from knowing the truth."[30] It's the same old glass houses argument of W. L. George. And it's the same old sleight of mind that rationalizes revealing more information about those lower on the hierarchy while concealing it about those higher: do you think you'd be allowed to peek at a CEO's or a politician's black box recorder anytime you wanted?

Meanwhile, it's not just black boxes that store data on how fast you're driving: autos can be equipped, or planted, with GPS devices. Cops put GPS devices on cars to follow suspects. Stalkers put GPS devices on cars to follow victims. "Experts who train victims' advocates, law enforcement and prosecutors recommend that you check underneath the hood of your car and look for suspicious-looking parts."[31] But you don't need stalkers or cops to install your GPS for you. There are people who do it themselves. Some of the more expensive cars now include "Onboard Assistance" GPS systems so that you know where you've been. And foreign business executives and diplomats, both at risk for kidnapping, are being urged to use implanted chips and GPS devices to prepare for a possible rescue.[32]

Of course letting everyone know where everyone else is isn't the goal here. The cells of the Panopticon must always be lit, while the guard stations must always be dark. The military, which developed GPS, has the ability to turn off GPS signals. They call it "selective availability" to "degrade the quality of GPS available to civilians." That ability has been denied by a May 2000 executive order that states the military has to pinpoint areas if it decides to degrade GPS quality. So far the news reports that only unsavory characters (such as the Taliban) have been denied the benefits of GPS technology.[33]

The point is that we should never be deceived that technologies are neutral. They are controlled by those in power, which means those in power have the ability to gain access to information about those under their power, whether or not those under their power desire this information known. This information is then used to reward those the powerful choose to reward, and to harm those they wish to harm. At the same time, those not in power do not have access to the same sort of information about those in power. Within this rubric, information, like power, is a one-way street, and a dead-end one, at that.

Nothing to Fear

We are not content with negative obedience, nor even with the most abject submission. When finally you surrender to us, it must be of your own free will. We do not destroy the heretic because he resists us; so long as he resists we never destroy him. We convert him, we capture his inner mind, we reshape him. We burn all evil and all illusion out of him; we bring him over to our side, not in appearance, but genuinely, heart and soul. We make him one of ourselves before we kill him.

George Orwell

Economics and politics have become inextricable. One could argue that they always have been, that, as philosopher John Locke put it, "Government has no other end but the preservation of property."[1] James Madison seemed to agree, as he insisted that the main goal of the American political system was "to protect the minority of the opulent against the majority."[2] Adam Smith, godfather of modern economics and darling of those who are currently killing the planet, agreed as well, stating, "Civil government . . . is in reality instituted for the defense of the rich against the poor, of those who have some property against those who have none at all."[3]

Corporations and governments are two wings of bureaucracy, which pursues technological progress as an end in itself. Bureaucracy limits choices of behavior, increases the knowledge and information gap between the haves and have-nots, lowers the average level of public understanding, increases instability in economic markets, and increases unaccountability (and eventually mistrust and resistance) in the political realm. More instability encourages more surveillance and attempts to control, which encourages more mistrust and instability. But few see any alternatives to technological advance, or question the futile strategy of protecting legal rights to privacy with yet another bureaucracy.

"Privacy" is nothing but an illusory freedom when one's economic and political choices, and even one's worldview and one's identity, are constricted by the corporate means of production, the hijacking of political representation, and the propaganda of the mass media. The right to privacy is a perfect cookie to offer the servant whose freedom (time, place, lifestyle, view) is carefully circumscribed. Prisoners shall be granted several hours a day to be left alone to choose among an array of frozen dinners and television shows. If prisoners complain about their frozen dinners or do not watch their television shows, they will no longer be left alone. But if they sufficiently appreciate these granted freedoms, they shall also be let out of work a week or two every year to choose between several theme parks and several automobile tours of the Interstate Highway System.

Pretend you're John Ashcroft. Yes, *that* John Ashcroft. General John Ashcroft (sorry, *Attorney* General John Ashcroft). Pretend, even more implausibly, that the reality of being John Ashcroft does not drive you to kill yourself as quickly as possible. (Attorney General John Ashcroft responds, "That's what makes America great. There are some countries around the world where I could [and would] have you shot for saying that [in fact I'm considering sending you to one right now]. But here I merely put you under surveillance [for now]. You can say anything you want [for now], so long as what you say or do does not in any way impede economic production and doesn't interfere with my full and free exercise of power.") Pretend you lost a Senate race to a dead man. Pretend you draped a cover over the naked breasts of a statue of Miss Justice in the halls of the Justice Department (perhaps to further disembody justice). Pretend you want more power to intervene in what are called Arab hawala transactions (where cash is exchanged in an honor system) because people from Arabian countries have funneled money to terrorists. (No, silly, not the U.S. military: those are the terrorists *you* funnel money to through taxes. Ashcroft means the *other* terrorists, the ones he doesn't like [at least right now]). You want the capacity to monitor all financial transactions more closely. You want to be able to get business records without a court order. You want to track wireless communications with a roving warrant. And for good measure, because everyone knows drugs and terrorism go together like the FBI and CIA, you want to increase the sentences for drug kingpins to forty years in prison and $4 million in fines.[4] (And no, silly, not the CIA: Those are the drug kingpins *you* funnel money to through taxes. Ashcroft means the *other* drug kingpins, the ones he doesn't like [at least right now]).

What do you do?

Well, if you really *are* John Ashcroft, you take your show on the road. You tour the country promoting your latest act, in this case the otherwise unpopular Vital Interdiction of Criminal Terrorist Organizations Act (VICTORY Act, a.k.a. PATRIOT II Act). You remember that in your line of work, perception really does trump reality, and you remember (as if you could ever forget) that people obsessed with fear are obsessed with security. With one hand you scare them and at the same time provide ways to make their fears abate (but just for a little while). With the other, of course, you draw a noose ever tighter around their necks.

From the CIA interrogation manual: "1. The more completely the place of confinement eliminates sensory stimuli, the more rapidly and deeply will the interrogatee be affected. Results produced only after weeks or months of imprisonment in an ordinary cell can be duplicated in hours or days in a cell which has no light (or weak artificial light which never varies), which is sound-proofed, in which odors are eliminated, etc. An environment still more subject to control, such as water-tank or iron lung, is even more effective. 2. An early effect of such an environment is anxiety. How soon it appears and how strong it is depends upon the psychological characteristics of the individual. 3. The interrogator can benefit from the subject's anxiety. As the interrogator becomes linked in the subject's mind with the reward of lessened anxiety, human contact, and meaningful activity, and thus with providing relief for growing discomfort, the questioner assumes a benevolent role. 4. The deprivation of stimuli induces regression by depriving the subject's mind of contact with an outer world and thus forcing it in upon itself. At the same time, the calculated provision of stimuli during interrogation tends to make the regressed subject view the interrogator as a father-figure. The result, normally, is a strengthening of the subject's tendencies toward compliance."[5]

Those in power like to tell the rest of us that they are instituting new surveillance regulations and technologies to protect us from (foreign) terrorists. But that's simply not true. U.S. air passengers were being screened well before 911. In 1998, the first version of the Computer Assisted Passenger Prescreening System (CAPPS) was alerting authorities to suspect passengers, such as those who bought one-way tickets with cash just before a flight.

After 911, Congress called for a beefed-up system, so the U.S. Transportation Security Agency (TSA) hired the military contractor Lockheed Martin (in an open-ended contract, of course) to build CAPPS-2 using "commercial data warehouses containing names, telephone numbers, former addresses, financial details and other information about virtually every adult American."[6] Under the new system, all potential airline passengers will undergo background checks after they book flights. The TSA won't reveal exactly what parts of your life will be investigated, but your name, address, birth date, and phone number will be checked against your credit reports, banking records, and criminal records.

CAPPS-2 will color-code every passenger, green, yellow, or red. The airlines will be given the ratings and decide "whether a passenger should be allowed to board or be subjected to additional questioning."[7] Reds will be barred from flying and referred to police.[8] CAPPS-2 is being designed to store information about people labeled yellow or red for fifty years.[9]

Plans are already in place to extend these data surveillance regimes to other forms of domestic travel, such as trains, buses, and even drivers' licenses.

In a classic example of data creep, CAPPS-2 information "could also be shared with other government agencies at the federal, state and local levels, as well as with intelligence agencies such as the CIA and with foreign governments and international agencies—all of which could use those designations for many purposes, including employment decisions and the granting of government benefits."[10] And in fact, the creep was already included in the testing. In 2002, "JetBlue Airways secretly turned over data about 1.5 million of its passengers to a company called Torch Concepts, under contract with the Department of Defense. Torch Concepts merged this data with Social Security numbers, home addresses, income levels and automobile records that it purchased from another company, Axciom Corp. All this was to test an automatic profiling system to give each person a terrorist threat ranking."[11]

In March 2003, TSA and Delta Airlines tested CAPPS-2 at three undisclosed airports. Shortly thereafter, the threat of lawsuits against the government[12] and boycotts against Delta Airlines[13] motivated TSA bureaucrats to try to mollify public concerns without making any significant change by announcing that CAPPS-2 would be "postponed," but the TSA's schedule for implementing the system is on track.[14]

Sure enough, by September 2003 (ironically, on the eve of the anniversary of the 911 hijackings) the U.S. government trotted out another announcement about CAPPS-2. Up to 8 percent of passengers who board the twenty to thirty thousand daily flights in the U.S. will be coded yellow and undergo additional screening at the checkpoint. One to two percent will be labeled red and denied boarding (for those of you doing the math, that's a couple of people from every large flight). In case you suspected only hard-core terrorists would be weeded out, a TSA spokesman declared that "not only should we keep passengers from sitting next to a terrorist, we should keep them from sitting next to wanted ax murderers"[15] (not that I've seen many people carrying axes on planes: my experience is that ax murderers generally prefer to travel by Greyhound).

Weeding out the many ax murderers among us won't be easy. It'll take more than color-coding at the airport. That's why CAPPS-2 will have airlines submit to the TSA the following information about everyone who books a flight: name, home address, telephone number, date of birth, and travel itinerary. "If the computer system identifies [*sic*] a threat, the TSA will notify federal or local law enforcement authorities."[16] This means you could be arrested before you ever leave home, merely for booking a flight. Now *that's* security.

If you're fretting about privacy, rest assured that the civil liberties folks are on the scene, protesting, rightly enough, about CAPPS-2, "You could be falsely arrested. You could be delayed. You could lose your ability to travel." You can rest just as assured that these protesters will limit themselves to demanding minor adjustments to the form of the Panopticon, and will fail to protest, question, or perhaps even perceive the Panopticon itself. And you can rest even *more* assured that the government will continue to refine and ratchet up its abilities to gain information about, and thus power over, those it purports to serve. A government spokesperson said, "Given the dynamic nature of the threat we deal with, it would be impossible to predict when the work would be finished [on air security]. We don't think it will ever end."[17]

It won't be all government bureaucrats doing the surveillance and sorting. It'll be corporate bureaucrats as well. By the end of 2004, airports will be replacing the federal force with private screeners.[18]

Not everyone is opposed to the new improved CAPPS system; a reporter for *Wired* was able to find at least one woman who said, "Whatever works, hon. I'm willing to give up a little privacy so that we're never attacked again. Besides I have nothing to hide."[19] And in March 2004 the airlines tried to distance themselves from citizen umbrage by adopting a set of procedures that would only collect personal information for aviation security purposes, inform passengers of its collection, keep the information secure, dispose of it after the passenger's trip, and observe the data collections protocols of other nations.[20]

So long as we do what they tell us, we have nothing to hide, and nothing to fear.

So long as we do what they tell us, we have nothing to hide, and nothing to fear.

So long as we do what they tell us, we have nothing to hide, and nothing to fear.

Just keep telling yourself that.

More from the manual:
"It has been plausibly suggested that, whereas pain inflicted on a person from outside himself may actually focus or intensify his will to resist, his resistance is likelier to be sapped by pain which

he seems to inflict upon himself. In the simple torture situation the contest is one between the individual and his tormentor. . . . When the individual is told to stand at attention for long periods, an intervening factor is introduced. The immediate source of pain is not the interrogator but the victim himself. The motivational strength of the individual is likely to exhaust itself in this internal encounter. . . . As long as the subject remains standing, he is attributing to his captor the power to do something worse to him, but there is actually no showdown of the ability of the interrogator to do so."[21]

So long as we do what they tell us, we have nothing to hide, and nothing to fear.

For the most part our imprisonment is not in a building, but rather in a state of fear. Michel Foucault knew that the "ultimate purpose of the panopticon is not to imprison the body, but to induce in the inmate [the student, the customer, the citizen, the human being] a state of conscious and permanent visibility that assures the automatic functioning of power." To punish less, perhaps; but certainly to punish better.[22]

Surveillance is more than spying, and it's more than torturing people until they "regress." Emile Durkheim and others have redefined surveillance as the "gathering of information about and the supervision of subject populations in organizations," and they regard surveillance and bureaucracy as a "rational response to the size and complexity of administrative tasks posed by science and technology."[23] In his book *The End of Privacy,* Reg Whitaker defines surveillance as the collection and analysis of information for the purpose of control. In either case, the surveillance is done by the main institutions of modern society: the central state and the business corporation.

If you recall, early on we quoted Foucault as saying that the Panopticon is "an important mechanism, for it automizes and disindividualizes power. Power has its principle not so much in a person as in a certain . . . arrangement whose internal mechanisms produce the relation in which individuals are caught up. . . . There is a machinery. . . . Consequently, it does not matter who exercises power."[24]

Enter the bureaucrats.

Weber wrote that bureaucracy is "fundamentally domination through knowledge. This is the feature which makes it specifically rational."[25] But this statement standing alone misleads on a couple of major counts. The first is that it's not domination *merely* through knowledge. Beneath it all, when the bureaucracy does not suffice, the state can use its monopoly on physical force. Weber's identification of discipline as the bureaucrats' second characteristic, and comparison of bureaucracy to a machine without will or mood, can also be misleading.[26] Bureaucrats may be narrowly rational, but their rational discipline serves structures and goals that are largely ignored by everyone. And while their jobs may depend on stifling their own will and mood, bureaucrats are fundamentally instruments of the will and mood of those in power. Bureaucrats' discipline is not self-discipline. Power does not discipline itself; it disciplines the populace through the administrations of bureaucrats.

When modern authorities say, as they often do, that "we are a nation of laws," they are pointing not at themselves but at those over whom they rule. Everyone knows that laws do not apply to those who make and implement them. Within our culture they never have. As B. Traven, author of *The Treasure of the Sierra Madre* and other books, wrote, "'Law and Order' means: The protection of property, the protection of capital. 'Law and Order' means: To protect the capitalists so that they can in a lawful and orderly fashion impoverish those who want to eat. The shopkeeper cries for 'law and order.' The rulers, and those who want to rule cry for 'law and order.'"[27] When those in power emphasize the rule of law, they are pointing to the narrow rational disciplining of the masses, and away from the need for self-discipline. We would only substitute our bureaucrat for Traven's capitalist.

Anatole France wrote, "The law, in its majestic equality, forbids the rich as well as the poor to sleep under bridges, to beg in the streets, and to steal bread."[28] France's point was clear. The poor are not to be free or independent; the rich do not have to beg, as they control the levers of power, which are the privatization of profits and the externalization of costs, or the taking for themselves of whatever material benefits the machine provides, while forcing others (including especially the nonhuman world) to suffer the material consequences.

The implications of the rational disciplinarian society are not hidden, but ignored. How hidden were the Nazi extermination camps? How hidden are clearcuts? How hidden is a planet being converted into cash? Nothing's hidden, but we don't see it, and if we're asked, there are countless rationales for continuing to not pay attention.

Rationality exists to explain and justify, and it ignores what it cannot explain and justify. The purpose of rationality is the "logical and efficient" pursuit of a goal that is determined by emotion, either an unconscious compulsion or a conscious act of will. Once the goal is set, the rational bureaucrat's only duty is to fulfill it. This is as true, once again, of killing Jews as it is of killing Indians as it is of killing forests as it is of killing the oceans.

Last year George was at the university doing research on bureaucracy and power. When he tried to use a database on the university's computer system, the computer "told him" he had to create an account to use the database. So he identified himself by his alumni number and authenticated his identity by supplying his social security number. Now he had an account. But he was denied access to the database he wanted, on the grounds that he wasn't an authorized account holder because he was an alumnus, not a current university student. He'd been duly identified and sorted and excluded according to the rules of the Panopticon, the impersonal machine that has as its function to identify and sort and grant or deny goods and services. He was a bit annoyed he'd been lured by the siren of free unlimited computerized information into spending half an hour fiddling with computer accounts but was relieved to go back to the library shelves to read good old-fashioned printed books. After all, he thought, Marx, Weber, Schumpeter, Ellul, Foucault, and other scholars had articulated much of what we need to know about bureaucracy and power without using computers—so he figured he ought to be able to learn what they had to teach him without resorting to infernal machines.

But last year's attempt to join the university's computerized system is still costing George time and effort. This morning he received an email message from the university advising him to log on to the system to make sure his account was in good working order. Duly following orders from a machine he hadn't been able to use a year earlier, he tried to log on, but of course had forgotten his password. He sent an email message to an anonymous support staff, saying he didn't really want to go through the hassle of reactivating his account. Instead, what he really wanted, since he wasn't allowed to use the system anyway, was to *deactivate* it (cancel? expunge? eliminate? kill that part of his identity?). He was told that no one could deactivate his account but him. But he can't access the account without proving he is himself. The solution he was offered? Continue to work with the computerized interface (the rules of which he does not understand, and which the system's engineers seem unable to explain), or physically come over to the campus and prove that he's himself, at which point a bureaucrat may be able to reactivate his account, and then he can deactivate it.

Identification, authentication, separation, elimination.

The panoptic sort.

You go to work. You have a very busy day ahead of you. You always do. Your work is very important. You are a transportation engineer. Your job is to manage a crucial sector of the economy. It's complicated. You love telling your friends—as they struggle to keep their eyes from glazing—about the difficulties of making sure the trains run on time, and of making sure resource A arrives at factory B before the factory runs out of raw materials, and that the train used to carry resource A never leaves factory B empty. It needs to carry product C to endpoint consumers. "That's you," you like to tell your friends, who smile weakly and blink their eyes.

"Sometimes we have to use trucks," you continue, "if we have a sudden glut of resource A to move. Not just anyone can predict when more resource A will be released for us to use. That takes expertise and experience."

You finally notice that their attention is lagging, so you pull out your trump card. You say, "You do like your product C, don't you?"

"Oh, yes," they say, finally enthusiastic. "We like our product C very much."

This is how the world is destroyed. This is how atrocities are committed.

Resource A = Jews (each of whom had, before being sorted by the Panopticon, a life and personality and desires unique to that individual).

Factory B = Auschwitz, Treblinka, or any other death factory.

Product C = Eyeglasses, bales of human hair, soap made from the flesh of humans, and so on.

Or,

Resource A = Cows (each of whom had, before being sorted by the Panopticon, a life and personality and desires unique to that individual).

Factory B = A factory slaughterhouse or any other death factory.

Product C = Hamburgers, steaks, and other products made from the flesh of cows.

Or,

Resource A = Trees (each of whom had, before being sorted by the Panopticon, a life and personality and desires unique to that individual).

Factory B = A sawmill or any other death factory.

Product C = Toilet paper, chopsticks, newspaper, and other products made from the flesh of trees.

It doesn't matter what (or rather who) A is. A could be any living being. That is the strength of the machine: individuals do not matter. They never matter. All can be consumed. All can be converted to fuel for the machine.

It doesn't matter what kind of factory B is. B could be designed to convert any type of living beings into products, that is, to kill them. That is the strength of the machine: individuals do not matter. They never matter. All can be consumed. All can be converted to fuel for the machine.

It doesn't matter what kind of product C is. It could be anything from food (or in the case of most of what we eat, "food") to nuclear bombs. That is the strength of the machine: individual products do not matter. They never matter. All of life can be turned into generic products. Interchangeability is the key.

Standardization, utility, efficiency, interchangeability. How to destroy the world is as easy as A, B, C.

Let's say you're not a transportation engineer. Let's say you're appalled by the use of resource A at factory B to make product C. You confront the transportation engineer.

Chances are good this person won't know—or rather acknowl-edge—what you're talking about. "I just make the trains run on time," the engineer might say. "And that's a very difficult job. You wouldn't believe how much expertise and experience it takes."

You can already feel your eyes glazing.

Bureaucracy (I first typed Bureaucrazy) becomes the perfect screen to mask culpability. If everyone is a cog, who is account-able? Just as the scientists at vivisection labs need not hate ani-mals, so too janitors can feel unaccountable for the atrocities com-mitted by the bureaucracies of which they're a part. Similarly, mail clerks just carry mail. Operators just run telephones. (Both of these services are now run by machines, instead of people trained to be machines.) Scientists just do research. Managers just make schedules. CEOs just try to maintain solid revenue flows for (themselves and) shareholders. Shareholders just cash checks. Politicians just provide tax breaks and regulatory loopholes. Police just protect property.

If an individual were to burst into your home to beat and rape you, you could eliminate this threat by putting a bullet into the intruder's brain. But part of the brilliance of the machine struc-ture, of the corporate structure—which has been granted not only personhood but immortality (and immorality) by the govern-ment—is that accountability has been essentially eradicated. Hell, scratch *essentially*. The point of a limited liability corporation is to limit liability. Every cog in the machine does its part as the machine consumes the souls of living beings, as it consumes the material world. The machine has no brain into which we can fire a bullet. Whom or what do we shoot—figuratively or literally—to stop this relentless conversion of the living to the dead?

The historian Martine Kaluszynski wrote about the French gov-ernment at the end of the nineteenth century: "[I]ndustrial growth and urbanization had radically transformed the way of life and destabilized the existence of a significant proportion of the population. The response of the [government] was to extol the virtues of order, stability, and work, and it did all in its power to enforce respect for these values. . . . [A]nthropometry was not simply a new weapon in the armory of repression, but a revolutionary technique: it placed identity and identification at the heart of government policy, introducing a spirit and set of principles that still exist today."[29]

When she talks here of "anthropometry," remember that she's talking of those quaint anthropometric techniques of measuring heads and ears and arms. But she's dead-on when she says biometrics was used first as repression and then as basic government policy. That's how it usually goes. What one generation perceives as repression, the next accepts as a necessary part of a complex daily life. The panoptic sorting and disciplining of the population *is* the basic bureaucratic method of governance. And it is with us today more than ever.

This is the rational discipline of the modern system. Bureaucrats (and consumers, if they want to get the rewards and avoid the penalties) erase their will and emotions to serve a rational and disciplined machine. The cogs in the machine are rational, geared toward fulfilling certain narrow programmed ends, such as efficiency of production, maintaining high levels of consumption, securing the established socioeconomic order, and so on. The cogs are ever-dedicated to fragmenting community and disciplining themselves. But is the system itself rational? Is it disciplined? Or is it a self-fulfilling circular dead-end? Bureaucracy and its tools, science and technology, have become self-justifying ends in themselves. We know the rules (rules, rules, everywhere) and our duties: go to work, use our spare time to enjoy our work's fiscal benefits, and protect the system as if it were our home.

But what's the point?

. . . .

Well, we all know the answer to that. The point of the machine
is that it, like any machine, concentrates power (these days often
in the form of wealth). And being dependent on the system, the
cogs—that's generally you and me—also want to make their little
share, want to gain a little power for themselves. So liberals and
reactionaries alike invent and sell the tools of wealth-making and
power-wielding: airport security services, commercialization of
information, sorting each other into the various market sectors,
and so on.

The original Panopticon consisted of all sticks and no carrots,
which is an expensive and unstable way to maintain a system of
oppression. Bentham was aware of this, writing, "In stating what
this principle will do in promoting the progress of instruction in
every line, a word or two will be thought sufficient to state what
it will not do. It does give every degree of efficacy which can be
given to the influence of punishment and restraint. But it does
nothing towards correcting the oppressive influence of punishment and restraint, by the enlivening and invigorating influence
of reward."[30]

That's a problem. That's a big problem. Even though the
United States is a major exporter of such torture devices as shock
batons and "iron wreaths" (head screws progressively tightened
around the head or ears) and of course counts torture as a significant tool in its foreign and domestic policy toolbox, the application of these tools, for reasons of administrative and financial
ease, are normally limited to those who either can't be bought or
those the panoptic sort determines aren't worth bothering to buy.

Which means that most of those who *have* been bought, which
means most middle-class Americans and Europeans, which
means most who have slid down the right tubes and into the
proper bins of the panoptic sort, can feel and believe they aren't
targets for the security apparatus (but of course if they have any

fear of terrorists, hurricanes, disease, or crime, the agencies that provide them with "security" have some level of control over them). Those who have been sorted and sold can also say and believe, "Thank God (or providence) we do not live in the dreary societies of the past! Technological progress and democratic self-governance are the fruits of our hard work and virtue, and the average person now [or at least the average human white property owner in the USA] enjoys luxuries and comfort undreamt of by the kings and queens of old—and the new knowledge-based economy means work is easier and more pleasant than ever!"

It is no mistake that modern consumer society provides a corresponding universal, endless, no-escape environment and mind-state of greed and need to go along with its paired mind-state of fear. The shopping mall is the panoptic architecture of consumption. The hallways are endless. You get lost in the stores. The stores are designed to instill a timeless sense of distraction, enchantment, and addiction.

There are two basic ways those in power can convince us to give them information about us, corresponding to our greed and our fear: economic "benefits" and "security." Economic incentives include, among many other things, giving product rebates to consumers and making social security payments. Security incentives include protecting our bank accounts and keeping "undesirables" from entering "our" country. So bureaucracies, that is, governments and corporations, alternately give us the carrot of economic benefit and the stick of security. Depending on where you are in the panoptic sort, the same carrot-stick may oppress or bless you.

Bless comes from the Old English *bldsian,* to consecrate, which is derived from Germanic *bldan,* blood, to sprinkle with blood. Blessed are the meek for they shall inherit the earth, but not until they are washed in their own blood.

The Real World

Whoever uses machines does all his work like a machine. He who does his work like a machine grows a heart like a machine, and he who carries the heart of a machine in his breast loses his simplicity. It is not that I do not know of such things; I am ashamed to use them.

Chuang tzu

The global culture of consumption hasn't evolved out of some inherent human addiction to the ownership of "goods." Before human beings can be addicted to "things," they must be converted from human beings to consumers. This is often accomplished through the destruction of those humans still living in supportive communities interacting with their landbase. As we have seen time and again: the indigenous do not become civilized until their human and nonhuman communities have been decimated. Once communities are gone, conditions are ripe for the creation of consumers and mass commerce, which is driven by merchant greed as much as by consumer demand. If I can't live in a human community embedded in its natural surroundings, I may as well have a good laptop computer and a really groovy stereo. Greed is the "offensive" side of fear; it's the seeking of comfort. Security is the "defensive" side of fear; it's the protection of comfort. Money is the medium of exchange that comfort-seeking takes in the modern socioeconomic system. Wealth is the turning of earth, lives, labor, and so on into consumables and money. Endless profit-seeking turns the wheels of commerce. Some (Rand, RAND, and others) say that's the genius of the capitalist system. And it is.

• • • •

But it isn't. The first definition of *genius* in the dictionary is the attendant spirit of a person or a place, as in *djinn,* as in *genius loci,* the pervading spirit or tutelary deity of a place. But machines—including societies formed as machines, as in capitalism, as in industrialism, as in civilization—are placeless. And they are spiritless. Which means they can never have genius. They can have power. They can have organization. They can have information. They can have a sort of virtual intelligence. But uprooted from place and deprived of spirit, they can never have genius.

We spoke earlier of two forms of identity theft: the common usage of that phrase, the theft of one's financial information; and the more important form of identity theft, getting us to identify with the machine instead of our own bodies, our own lives, and our landbases. But an even more pernicious and destructive third form of identity theft pervades our culture. We not only rob ourselves of our identities, but we rob everyone else of theirs too.

Consider our conceit that nonhumans do not have lives all their own. Reading this book, did you blanch when I wrote of a dragonfly asking me to not yet take its picture? Did you think I was speaking metaphorically, or projecting? Perhaps you thought, *Nonsense, everyone knows that only humans have volition, only humans speak, only humans communicate, only humans can express that they do not wish to have their pictures taken* (not that, inside the Panopticon, refusing to have your picture taken means anything to those at the center, except, perhaps, that you have something to hide).

Do plants have identities? How about rocks, stars, rivers? How about watersheds? Do they have desires all their own? Do they make choices and have existences and needs that may have nothing to do with what you want (or say or think you want, or have been convinced by the machine you want, or that serve production, or that serve the deathly ends of the machine, insofar as there is a difference between these latter)? Does a rabbit in a laboratory have a reason to exist apart from having caustic chemicals put into its eyes? Does a chimpanzee in a laboratory have a reason to exist apart from being infected with HIV? Does a redwood tree have a reason to exist apart from being dollar bills on a stump? Does a river wish to be dammed? Does your landbase wish to be covered with asphalt?

Or more to the point, does it matter to you what the rabbit, chimpanzee, redwood, river, or landbase want?

In machine cultures, even the most open-minded of cogs perceive listening to the natural world as a metaphor, as opposed to the way the world really works. Indigenous people listen to their nonhuman (and human) neighbors (and to their own bodies[1]), knowing that their nonhuman neighbors are individuals with lives worthy of respect. They do not steal the identities of their nonhuman neighbors by pretending that these neighbors have no identities at all. They do not grant personhood to corporations while denying it to living beings who share 99 percent of their DNA.

I hate the word *resources*. Resources do not exist. Salmon are not a resource. They are salmon people. They are fish, with lives as distinct from each other's as mine is from my sister's. My dictionary defines resources as "a natural source of wealth or revenue," or "computable wealth." I don't think salmon would agree to define themselves as a source of revenue or computable wealth. Nor would redwoods, rabbits, or rivers. Nor would human beings, at least those who have not already bought into (been sold) the myth of the machine.

I hate the word *ecosystems*. Ecosystems do not exist. The hummingbird looking in through the window perhaps reminding me I have not filled their feeder is not part of a system. Nor is the phoebe who sits on a branch overlooking the pond. Nor are the redwood trees reflecting sunlight from their needles. Nor the bear I heard breaking branches last night as she ran through the forest.

Nor the slugs I see eating scat on the forest floor. *System:* a regularly interacting or interdependent group of *items* forming a unified whole; a group of interacting bodies under the influence of related forces (a gravitational system).

This is the language of the machine. This is not the language of relationship. This is not the language of individuals working together in communities. *Community:* an interacting population of various kinds of individuals in a common location; a group of people with a common characteristic or interest living together within a larger society.

The machine is a way of perceiving and being in the world. We may perceive that we are living inside of machines, but that is wrong. I do not live inside of a machine. I do not live inside a complicated system of flowing resources, *items* interacting under related forces. I only pretend I am a cog, a resource, inside this system, inside this machine. But the truth is that whether I acknowledge it or not, I am an animal member of a vibrant and living community composed of people—some with wings, some with leaves or needles, some with hearts of stone, some who flow to the sea—all conducting their lives according to their own desires, intents, and volition.[2]

I am not a cog. I will never be a cog. I refuse to be a cog. I am a living being.

The University of Southern California (USC) is touting a machine able to amplify a sound 1,000 times and distinguish it from the "hubbub heard in bus stations, theater lobbies and cocktail parties"—even if the hubbub is 560 times louder than the target sound. No wonder the inventors claim the machine is superhuman. But true to cutting-edge science, the machine was based on "neural nets [that] are computing devices that mimic the way brains process information."[3]

Well, that's nonsense. The human brain doesn't use a central processing unit that requires programming; it uses a neural network that is able to learn from experience over time. The USC network uses eleven neurons with thirty connections. Biological systems—also known as living beings—have millions or billions of neurons. How many connections must there be in a healthy natural (human or nonhuman) community?

I have been thinking again about money, and I fear the preachers of my childhood were probably right: it is the root of evil, or at the very least much of it. But they were also wrong. Because money is not a root of anything. First, roots are part of living plants, and money is not alive. Second, money is nothing. A dollar has no significance on its own. I cannot eat it, and it does not provide all that much heat when burned. I cannot build a shelter from it. But within this shared delusion we call civilization it stands for power. And the urge for power over others is the source of much evil.

To a politician, money is the weapon for wielding power. For the businessman, power is the weapon for making more money. The delusion that money is worth anything, combined with this homicidal and suicidal urge for power, is strangling life on this planet. Few indeed are the scientists, politicans, and business-people (or citizens or consumers, for that matter) who can keep their logical or ethical perspective once the power and money start flowing.

It is worth noting that those living in healthy communities—not that many of us would recognize health or a community at this point—do not seem to have such trouble keeping logical or ethical perspective in the face of money and power.[4] This is one reason those in power must always destroy healthy communities.

Money

The coalition of money power with political power was one of the decisive marks of monarchic or despotic absolutism; and the more dependent the military machine became on technical inventions and mass production of weapons, the greater the immediate profits to the national economic system—even though in the long run succeeding generations would find these putative gains offset by the cost of reparations, repairs, and replacements, to say nothing of human wretchedness. Though the moral onus for promoting war has made the munitions manufacturers the scapegoats, the fact is that the paper-profits of war equally enrich every other part of the national economy, even agriculture; for war, with its unparalleled consumption of goods, and its unparalleled wastes, temporarily overcomes the chronic defect of an expanding technology—"overproduction." War, by restoring scarcity, is necessary on classic capitalist terms to ensure profit.

Lewis Mumford

Panoptic sorting is a necessary cost of doing business in a bureaucratic industrial economy. As Jane Caplan and John Torpey point out, "private economic and commercial activities would . . . grind to a halt unless companies had the ability to identify and track individuals as property owners, employees, business partners, and customers."[1] But the Panopticon is not just an overhead cost. The goods and services that are required to operate the Panopticon itself are a tremendous source of revenue for the modern economy. A few examples:

- Adding computer chips to all new passports authenticating the bearer's face and fingerprints would cost $100 million a year.[2]
- Setting up a national identity card system in Canada would cost $3 to $5 billion.[3]
- In 2002, the U.S. Congress passed legislation providing $900 million over three years for research and development to "secure cyberspace."[4]
- The U.S. government granted $1.9 million to the National Sheriffs' Association to double the number of Neighborhood Watch groups to fifteen thousand nationwide. President Bush requested $560 million in new funds for the 2003 fiscal year to support his USA Freedom Corps, which oversees several surveillance programs, including Neighborhood Watch and Operation TIPS.[5]
- In the 1990s, the U.S. Central Intelligence Agency gave Vladimiro Montesinos, the former head of the Peruvian National Intelligence Service, at least $10 million in cash, on top of surveillance equipment. (This was a drop in the bucket: Montesinos deposited at least $264 million in Swiss, U.S., and Cayman Islands bank accounts.[6]) All of this is chump change compared to the billions being spent to "fight drugs" in Colombia.
- In the winter of 2002–3, the U.S. Congress appropriated $54 million in tax money for the infamous (and supposedly suspended) Total Information Awareness program.[7]
- Macy's department store spends $28 million a year on security.[8]
- The U.S. Army has dedicated $50 million for MIT's new Institute for Soldier Nanotechnologies. Patriotic corporations, including Dow Corning, DuPont, and Raytheon, have pledged another $40 million so far. These investors' confidence is well-placed; during World War II, MIT's

Radiation Laboratory built $1.5 billion worth of radar systems.[9]

- California's Anti-Terrorism Information Center is getting more than $6 million a year in state funds,[10] this in a state where school budgets are being gutted.
- U.S. government agencies spent more than $50 million on camera surveillance technology in the five years preceeding 911, much of it for face recognition.[11]
- Some of the government's electronic surveillance programs are coordinated by the FBI's Technical Support Center, which is scheduled to receive $200 million a year.[12]
- More than 72 million smart cards (credit-card-sized plastic cards with an embedded computer chip) were sold in the United States and Canada in 2002; the global smart card market is currently worth $3.5 billion a year.[13]
- The U.S. Department of Homeland Security (DHS), which has 180,000 employees, hundreds of corporate contractors, and an annual budget of $37 billion, plans to spend more than $800 million on computer and information analysis upgrades in 2004.[14] The General Accounting Office warns that DHS's spending estimates are too conservative.[15]
- Seven hundred corporations and more than two hundred universities in Europe, the United States, and Japan invested more than $3 billion in nanotechnology in 2003. The U.S. government spent $2 billion on nanotech betweeen 2000 and 2003, and the European Union plans to spend $1 billion from 2002 to 2005.[16] That's a small percentage of what the market for nanotechnology is worth: $45 billion in 2003, and predicted to be $1 trillion by 2010.[17]
- More than half of U.S. corporations monitor their employees' e-mail; they spent $2.7 billion to do it in 2003 and will spend more than $6 billion per year by 2007.[18]

- The Pentagon estimates that there are forty million surveillance cameras in use worldwide and expects three hundred million to be installed by 2005.[19]
- In 1999, the U.S. government spent $22 billion on "civil and criminal justice." States spent $50 billion, counties spent $35 billion, and cities and towns spent $39 billion.[20] That's more than $146 billion, or $521 for every man, woman, and child in the country.
- Somewhere between $20 and $35 billion is spent every year to keep people in jails and prisons, and the prison system has more than half a million full-time employees.[21] Per-capita spending on prisons tripled between 1980 and 2000, from $44 to $129—but the prison population quadrupled over that time, to two million, meaning (hooray) that the system has become more efficient![22] More than three thousand new prisons were built in the 1990s at a cost of $27 billion.[23] One private "corrections" corporation, UNICOR, had 2002 revenues of more than $716 million.[24]
- Somewhere between $3 and $10 billion will be spent on an automated system to track foreigners entering the United States. U.S. consular officials abroad will fingerprint and photograph visa applicants in their home countries and check their identity against terrorist watch lists and criminal databases. U.S. border agents will scan travelers' index fingers to match them against their visa documents. A database will automatically alert the government to individuals whose visas have expired.[25]

From war to health, from airports to workplace e-mail, there's gold in them thar security scams. Check out the breathless opportunities offered in some of the past issues of *Biometrics Market*

Intelligence: Biometrics and the Government Part I: Waiting for The Windfall; Biometrics and Trusted Identity: Combating Identity Theft—Cashing in on Identity Theft; Travel and Transportation: Is Air Travel the Killer App?; Market Roundup: Vendor Success Stories Agro-Terrorism.[26]

Only those who still perceive governments as serving individuals and communities as opposed to overseeing the privatization of profits and the externalization of costs will be surprised that public "security" and private profit are inseparable. Around the world, welfare agencies, police, prisons, energy companies, and transportation utilities, along with the information they hold, are being "privatized," that is, sold at grossly reduced prices by public agencies to private corporations. Perhaps it's shocking to some that ChoicePoint and Axciom and other corporations—not the big brother feds—are now the primary sources of information on individuals. But the commercialization of detailed information about the public isn't new. Public opinion polls soliciting the "opinions" of citizens evolved from market surveys. Primitive machineries for panoptic sorting such as fingerprinting and the Hollerith card were, from the beginning, joint ventures between government and corporate bureaucracies. Don't be surprised. This is how the machine has always worked. You'd only forgotten for a time.

We spoke earlier of RFID chips. We want to emphasize that they are not science fiction. They are not something that may happen sometime far in the future. They are happening right now.

Beginning in 2004, Goodyear Tire and Rubber will install RFID chips in its tires. Michelin will follow suit in 2005. Tire type and size, serial number, manufacturing date, and other information can be stored on a chip connected to two filaments that serve as antenna. The company says if RFID chips had been in the Bridgestone/Firestone tires that caused fatal roll-over accidents in

Ford Explorer SUVs a few years ago, it could have identified exactly which tires caused the accidents. Eventually, RFIDs will be "smart enough" to "tell" vehicle owners and drivers "if the tires are properly inflated, overheated, overloaded, or if tire tread is dangerously worn."[27]

Unsaid, of course, is that RFID chips will also allow vehicles to be tracked automatically. Currently "passive" RFID tags have a range somewhere between an inch and forty feet. "Active" RFID tags have a range of miles. Current tire RFIDs are designed to transmit the chips' data to a receiver a mere thirty inches away from the tires, but what will the eventual range be? Will RFID someday be like GPS, able to pinpoint its bearer's location from any distance?

Procter & Gamble (P&G) plans to have RFID tags on consumer products by 2005. P&G spokeswoman Jeannie Tharrington claims that the tags "allow us to see what consumers are purchasing so that we can adjust to meet consumer demand more accurately" and promises that P&G isn't going to abuse the RFID data because "privacy is very important to us."[28]

There are lots of other uses for RFID tags, but none of them need concern you regarding your privacy. Your refrigerator could report its contents to the supermarket for restocking, and your television could be programmed to broadcast commercials based on the contents of your larder.[29] Why watch commercials that aren't relevant to you? Or, feeling ill? Whatever ails you, there's probably a pill for it by now, and those pills could be tagged with Auto-ID, so they could be remotely monitored. If you forget to take them, you could be reminded. It's all very convenient.[30] One should never forget to take one's soma.

The U.S. Army wants RFID chips too and is testing a $75,000, three-axle, six-wheel-drive Smart Truck to roll on RFID-equipped tires. Smart Truck is capable of carrying five tons, including forty-eight laser-guided Spike missiles and a Pointer.

In case you aren't monitoring the U.S. Army Automotive Center's progress, let us tell you that the Pointer is a remote-controlled unmanned aerial vehicle (UAV) that can stay in the air for five hours surveying a ten-square-mile area. That area could be anything from "enemy territory" to your backyard. The Smart Truck isn't just for war. It's touted to be "everything to all governmental services." According to one Army spokesman, the use to which any individual truck will be put "all depends on what is needed to do what is needed to be done."[31]

Gillette, Wal-Mart, and Tesco are already installing shelves that can read RFID radio waves embedded in shavers and related products,[32] and Gillette announced it would be buying five hundred million RFID tags from Alien Technology Corporation of Morgan Hill, California.[33] (Alien technology!) "Gillette has dismissed assertions by privacy groups that the company plans to use smart tags in its products to track and photograph shoppers." They'll leave that to others. The UK supermarket chain Tesco, where Gillette tested the RFID-equipped products, admitted it was taking photos of shoppers who picked Gillette razor blades off the shelves, or who left the store with tagged products. A Tesco rep said they were "just looking at the benefits. It is blue sky stuff. The camera use was a side project to look at the security benefit."[34] Function creep.

Already, the Dulles Toll Road lets cars with RFID chips "zip by in the fast lane." Mobil gas stations allow SpeedPass users with RFID cards to pay at the pump.

If this isn't scary enough, the chips have been developed by a corporation called the Matrics company. And who is behind the Matrics company's development of the RFID chip? National Security Agency veterans William Bandy and Michael Arneson. The Carlyle Group. Polaris Venture Partners. The Washington Dinner Club. The Women's Growth Capital Fund.[35]

Do you feel the net tightening around you?

Repeat after me: so long as we do what they tell us, we have nothing to hide, and nothing to fear.

In March 2003, clothing manufacturer Benetton announced it would put RFID chips in the labels of fifteen million garments sold at its five thousand stores worldwide. The chips would be used to track pieces of clothing from manufacture to sale (and, we can only presume, as technological advances allow, from there to closet, and everywhere else the clothes—and you—go). Responding to the threat of boycotts, in April 2003, Benetton announced it had backed away from the plan, and that none of its clothing contained the chips, but reserved the right to implement it in the future.[36] The upscale designer clothier Prada isn't so backward. RFID chips are used in its Manhattan store to "give salespeople up-to-date inventory and customer information." Customer. Yes, that's you. Salespeople are also equipped with handheld devices that control video displays throughout the store. Any article of clothing a customer tries on is automatically registered and displayed on a closet touch screen and stored in an Internet account that "the customer can refer to later." (Is anyone else watching?) Dressing rooms are equipped with a video-based "magic mirror" that shows the customer's back (no other angles?) "as well as a delayed replay of the customer turning around." (How delayed?)[37]

Of course, if you're not shopping at upscale stores and building a virtual wardrobe online, the video profiling might mean something else to you. Lawsuits have been filed against mainstream department stores, including Macy's, Dillard's, and J.C. Penney, for racial profiling. The stores say overzealous employees, and not the technology or the store management, are to blame.[38]

Chris McGoey of McGoey Security Consulting, who sometimes goes by the name The Crime Doctor, says he is often asked if stores use profiling tactics to identify suspect customers. "The answer is undeniably, yes. The concept of shoplifter profiling is a proven loss prevention tool and is currently being practiced in most major retail stores by trained loss prevention or security staff. Does that seem shocking? It shouldn't, as long as it doesn't include the discriminatory practice of *focusing on the race of the customer alone*. Profiling is used everyday as a method for quickly focusing in on a person, a product line or a section of a store most likely to contribute to shoplifting. All investigative agencies including the police, FBI, and others have used profiling as a tool to narrow the field of possible suspects. Why shouldn't retail store security be able to do the same? Store and customer profiles are developed during day-to-day operation and by collecting and analyzing inventory data. This data provides both a quantitative and a qualitative basis for determining where, when, how, and by whom shoplifting is likely to occur in the future."[39]

The point seems to be, if cops do it, why shouldn't store clerks? It's scientific (that is, quantified), so it's got to be good. McGoey declares that monitoring customers is "a matter of economic survival," which in this rubric means unavoidable. The challenge for McGoey isn't ethical or even merely technical; it's that profiling customers is "bad public relations if done crudely." So respecting the customer's modesty is one way to practice good public relations: "Customer surveillance is limited to the public areas where there is no expectation of privacy as opposed to inside fitting rooms and restrooms that are considered private areas." But the real standard to keep in McGoey's profession is secrecy: "Knowing that you are under surveillance is an uneasy feeling. No one likes being watched and being made to feel like you're not trustworthy. However, if trained professionals do the surveillance properly, most people will never realize they were observed while shopping."[40]

The core of the Panopticon: surveillance must be always possible, never verifiable.

If you have nothing to hide, why should you care if McGoey is watching you?

But McGoey is still practicing surveillance the old-fashioned way: the visual monitoring of selective suspects. In the brave new world, neither visuals nor racial discrimination are necessary; they are not only bad public relations, but too slow and error-prone. Why bother with labor-intensive spying and those pesky racial concerns? *Everyone* can now be implanted with miniature identification chips. Authorized dealers are already available to you in Arizona, Florida, Louisiana, Maryland, New York, South Carolina, Texas, and Virginia. Are you afraid your child will be kidnapped? Got a senile relative needs watching? Got an unruly child? Got a problem teenager? A continent full of free human beings? DigitalAngels, Verichipd, and other "technology that cares" is already here. The size of a penpoint, it doesn't hurt when it's installed. (Hell, maybe the person getting it won't even know!) It doesn't run on batteries, so it's good for up to twenty years. It only costs $10 a month to register your chip, so that you can be authenticated by anyone (anything) scanning your body. And the monthly charge can be billed to your credit card. "Get chipped!"[41] Thousands of teenagers have already volunteered to be chipped.[42]

They obviously have nothing to hide. Good kids, each and every one of them. They'll make fine cogs.

We know this.

The basic panoptic promise is security, either for the wardens or for the general population.

Military and commercial uses of RFID tags are scary. But even more troubling is that the world is being standardized to facilitate total technological penetration. This requires that the world be killed. As we see.

It also requires a lot of guards. Panopticon guards today face the same problem that the French, German, and Swiss watchers did when they took photographs of those who chose not to accept wages and mortgages. So many security risks to watch, so few guards to watch them. But that's exactly the promise of the RFID system. As one RFID critic put it, "The point of automatic identification, of course, is to take people out of the loop, to enable computers to gather information and act on it."[43] Just like the unmanned drones being used by the Pentagon. Just like screening at airports. Just like hiring and firing at work. Take the humanity, the unpredictable humans, out of the never-ending loop. Move toward a fully automated world.

Welcome to the machine.

One hope, I suppose, would be that the machine watchers are as incompetent as most of the human watchers. Providence help us all if the watchers ever become as efficient as they pretend to be. (And that's what they've been striving for from the beginning, to get rid of that pesky animal inefficiency.)

Years ago I knew someone who was on the run from the FBI. The feds wanted to question her to see if she knew anything about some animal liberations. (The case has since been resolved, so all you G-men can put down your walkie-talkies.) She got tired of running and settled down in her hometown. Trying to find her, the feds went to her parents, and to her former work places. No one told them where she was. But the feds needn't have bothered asking. When she returned home, she put her name in the telephone book.

Do these guys have a clue?

The FBI is developing a National Instant Check System, but how can they find terrorists when they can't even smell their own moles? The mission? Seek, Find, and Destroy Public Enemy Number One (the Terrorist). How? Give the FBI another $186 million to beef up their computers.[44]

The "intelligence community" bungled the warning signs leading up to 911. The Bush II administration disregarded or misrepresented the assessments of its own spies leading up to the invasion of Iraq. The FBI is accused of mishandling (by altering or corrupting) surveillance records of thousands of phone calls, faxes, and computer data obtained by the Foreign Intelligence Surveillance Court.[45]

How can we trust the Pentagon with intelligence matters when it can't even get strategic information out of naked tortured prisoners? "The combination of inexperienced interrogators and linguists, military bureaucracy and disagreements among language contractors, has hindered the effort of authorities to gather information from al Qaeda and Taliban fighters detained at a U.S. Navy base in Cuba."[46] In other words, they can't hire enough interrogators who speak the language of the "detainees." Beat them all you want, leave them in positions of crucifixion in the sun, make them kneel for days at a time, put them against walls and fire rubber bullets at them, but you still can't speak Pashtu and they can't speak English. What we have here is a failure to communicate.

Under oath, U.S. Attorney General John Ashcroft testified that "the top priority of the department [is] to protect America against acts of terrorism and to bring terrorists to justice." How will he do this?

He will continue to investigate who was behind the attacks on 911 (so long as these investigations do not lead anywhere that would make the present administration or its allies uncomfortable). He will get all 194 U.S. attorneys and 56 FBI field offices to aggressively implement the USA PATRIOT Act. He will hire 570 new Border Patrol agents. He will give the FBI $411 million for enhanced information technology projects, surveillance, intelligence, investigative and response capabilities, aviation, and security. He will scan and digitally store five million documents related to environmental and peace and social justice organizations—er, sorry, documents related to terrorist groups and organizations.

In addition, the Department of Justice's [*sic*] 2003 budget request included $348.3 million to "activate" four new prisons in West Virginia, Kentucky, and California, and to expand existing prisons in Illinois and Arizona.[47] These facilities would add over five thousand "critically needed beds to reduce overcrowding" caused by overdetention of rapists, killers, drug offenders, and foreigners, er, sorry, terrorists. The numbers bear out Ashcroft's critical need: there were thirty-two thousand federal detainees in 1996, up to sixty-seven thousand detainees in 2001.

How much does it cost to detain sixty-seven thousand people? The budget for federal prisons is over $4 billion per year. Of course, federal inmates are a tiny percentage of the people in prisons, county jails, and other "correctional facilities" in the United States: 2.1 million and growing at a rate of seven hundred new inmates every week.[48]

Ashcroft wants money for more than hard-ass detention facilities. He testified that cops, computers, and kids can help to protect U.S. borders, and his 2003 budget request included $15 million for "encouraging citizen participation in law enforcement, community safety and terrorism preparedness; and $60 million for the Boys and Girls Clubs."[49] On the street this would be called an intensive program of snitches. When snitches were recuited in the former Soviet Union and East Germany (German Democratic Republic, thank you very much), this was seen as part of the Big Brother state, the only possible way to keep an unpopular government in place. But we're different. Here it's necessary to keep us all secure.

In January 2003, the U.S. Congress blocked funding for the Pentagon's Total Information Awareness (TIA) project until the Pentagon showed it would protect the privacy of American citizens. In August, when the TIA tried to put forward its FutureMAP futures market, a venue for betting on terrorist

attacks (no, silly, not U.S.-sponsored terrorist attacks: *other* terrorist attacks), the project's head, hostages-for-weapons dealer
(and convicted felon) John Poindexter, was forced to resign. It
appeared that privacy concerns, civil liberties, common sense, and
basic ethics had gotten the upper hand. And they did. The TIA
project was given over to the lower hand. Not eliminated, exactly.
But hidden. As these things generally go.

Okay, so money for surveillance comes from taxpayers—in other
words, you pay to have someone watch over you—but where
does all the money *go?* One cause of American—and global
industrial—economic success has been the ability to turn every
problem to profit, and that's certainly true of terror—we mean
security. There are military contractors and other arms dealers.
There are private security guards and consultants and burglar
alarm companies (no more false alarms, you now get a micro-
phone in your house that allows the security company to listen to
what's going on in your living room in real time). There are contractors for the TIA.

There is Seisint, for example. With a reported $12 million in
federal funding, the private company Seisint is developing the
database that TIA was set up to create. The Multistate Anti-
Terrorism Information Exchange, dubbed the Matrix (showing
these guys can't recognize a film metaphor when it gives them a
martial arts kick to the face), was designed to tie together data
from local police records, state drivers' records, commercial credit
reports, county property records, and other sources, and to make
this enhanced data accessible to (so far) 135 local and state police
agencies in Alabama, Connecticut, Florida, Georgia, Kentucky,
Louisiana, Michigan, New York, Oregon, Pennsylvania, South
Carolina, Ohio, and Utah.

Seisint seemed a strong candidate for the complex data-
handling project, since it had already received numerous awards

for its work with local, state, and federal law enforcement agencies. Seisint had located missing children, helped with research on the World Trade Center hijackers, and routinely assisted in law enforcement, risk management, fraud detection, identity verification, and insurance investigations.[50]

From a gee whiz perspective, Seisint's hardware and software is pretty extraordinary. From the perspective of human beings who oppose the technologization and destruction of our lives and our landbases, it's pretty fucking scary. For instance, Seisint claims its computers can "instantly find the name and address of every brown-haired owner of a red Ford pickup truck in a 20-mile radius of a suspicious event."[51] While its Matrix system was being developed, representatives from Seisint met with Senator Bob Graham, who was head of the Senate Intelligence Committee. The Secret Service, the FBI, and the Immigration and Naturalization Service gave letters of commendation to Seisint (you can see them on Seisint's Web site). Seisint's amazing technology was demonstrated for U.S. Vice President Dick Cheney and Florida Governor Jeb Bush. It's no wonder the State of Florida and the U.S. Departments of Justice and Homeland Security saw fit to give Seisint funds and computer networking.

There's an even darker side to Seisint, however. Seisint was founded by Hank Asher. In the 1980s he smuggled illegal drugs into the U.S. from the Bahamas. He wasn't charged with any crimes; instead, he served as a snitch against other criminals, all the while providing police protection for yet other smuggling operations through his friendship with the head of Florida's Department of Law Enforcement. Asher was also a major contributor of political campaign funds, which always helps smooth things over, and which helped transform him from mere informant to high-tech data entrepreneur.[52] In 1999, the Drug Enforcement Administration and the FBI suspended contracts with another of Asher's companies "because of concerns about his past." That company, DBT Online, then bought out Asher for $147 million. And those letters of commendation that are on Seisint's Web site? The FBI and Secret Service have told reporters that letters like that are "routinely given as thank-you notes to hotels and other companies that help their agencies."[53] Still, Seisint executives include a former head of the Secret Service. One of Seisint's directors is a former president of the American Bar Association. Seisint's president said he'd never heard of Asher's drug activities; while these people know your hair color, address, the make, model, and color of your car, they say they know nothing about their founder's years of drug running. This would be unbelievable if we didn't understand the necessarily one-way nature of the panoptic gaze. And then there's the question of all those law enforcement agencies depending on the vast information capabilities of Seisint's computer databases to catch drug runners when the designer of these databases had protected members of this same group. We need not worry about the potential misuse or tainting of information, though, because as former American Bar Association president and current Seisint director Martha Barnett said, Asher is "a creative genius" at gathering information, and his past doesn't matter: "The truth is it's not about Hank Asher any more. He's come up with a terribly important product."[54]

Where else does the money go? Well, there's Victor Bout. A recent *New York Times* profile of Bout portrayed him as the world's biggest arms merchant.[55] A staffer at the U.S. National Security Council called Bout "brilliant" and said that "had he been dealing in legal commodities, he would have been considered one of the world's greatest businessmen." I'm afraid these fine distinctions are lost on me. According to the *Times,* Bout's aircraft also move gladiolas, frozen chickens, diamonds, mining equipment, UN peacekeepers, French soldiers, and African heads of state. It takes a community to raise a war zone.

But the article was wrong about Bout's status. Bout's sales are bush league compared to the Arms Dealers All Stars. The U.S. government, through its military agencies and corporate middlemen, has been and continues to be by far the largest purveyor of lethal technology. And behavior, for that matter.

There are numerous sectors in the terror economy, and they're making lots of money in the post-911 snake oil sideshow. Various of these industries are offsprings of fear and greed: the public's (supposed) fear of terrorists (what is a terrorist, anyway?) and the slick greed of those who run the corporate economy. These industries range from the surveillance and security industries to mass media and high-tech companies.

For example, there's L-3 Communications, a "leading merchant supplier of Intelligence, Surveillance and Reconnaissance (ISR) systems and products, secure communications systems and products, avionics and ocean products, training devices and services, microwave components and telemetry, instrumentation, space and navigation products." The company has contracts to install 425 bomb detection machines in U.S. airports. Sales were more than $320 million in 2002. L-3 has another $498 million contract to provide security at U.S. Air Force bases.[56] As a subcontractor to upgrade the Coast Guard's command and control and logistics system over the next twenty years, L-3 is anticipating $1.5 billion. But because domestic "security" isn't profitable enough, L-3 tries to "invest frugally" by getting technology through "partnerships with researchers at laboratories at universities like the Massachusetts Institute of Technology or Cornell, or it tries to commercialize technologies initially developed for military purposes."[57] So, taxpayers pay to develop the technologies. Then taxpayers pay for the installation of these technologies. Then these technologies are used to surveil the taxpayers. And the contractors, often former government officials, pocket the profits.

Not just purveyors of scary weapons and spooky surveillance technologies receive taxpayer assistance—so do old-fashioned information collectors. ChoicePoint has over ten (no wait, fourteen—no wait, twenty) billion pieces of data on Americans. It sells these data to insurance companies, to employers and landlords (to screen prospective workers and tenants), to the Boy Scouts (to screen volunteers), to marketers ("new customer acquisition programs"), and to cops. ChoicePoint's Bode Group performs DNA and other biometric identification analysis for federal and state agencies and commercial firms. ChoicePoint finds stolen cars and missing children. ChoicePoint can locate missing people who are owed inheritances or other assets, and can tell your creditors everything you own. If you're not sure who you are anymore, ChoicePoint can certify your birth or death certificates, and it can tell you whom you're married to. ChoicePoint was hired by Florida to erase the names of black citizens from voter roles. ChoicePoint is "the premier provider of decision-making intelligence to businesses and government."[58] Businesses and government. Businesses-and-government. Businessesandgovernment. ChoicePoint sells information to 40 percent of the one thousand largest corporations in the USA.[59] You'll be glad to know, "Protecting privacy is always a ChoicePoint priority."[60]

ChoicePoint was spun off from the credit agency Equifax in 1997. Equifax itself has records on four hundred million consumers, which it offers to, you guessed it, "financial services, retail, telecommunications/utilities, information technology, brokerage, insurance and business lending industries, and government."[61] It's not a monopoly, since ChoicePoint and Equifax have competitors. Two of the biggest are Experian and Trans Union.

Trans Union is owned by the Marmon Group, which holds over a hundred manufacturing and service companies with 550 facilities in forty countries. Marmon Group companies manufacture medical products and mining equipment, consumer products, transportation equipment, building products, and water-treatment products, besides providing marketing and distribution and consumer credit information. The Marmon Group is owned by Chicago's Pritzker family—the owners of the Hyatt hotel chain. Why are you worried about data creep? Totally secure, your credit records.[62]

Then there's Sybase, which sells software links, wireless applications, electronic banking services, and business intelligence software to governments, financial services, the health care industry, and telecom companies. Sybase sells Patriot Act–compliant software. "PATRIOTcompliance Solution integrates your existing customer and transaction information systems into a consolidated compliance system."[63] That helps "move data fluidly throughout the corporate infrastructure, regardless of the data type, the platform, the database, the application, or the vendor. This is Information Liquidity—transforming data into economic value. With Sybase, companies can attain maximum value from their data assets by getting the right information to the right people at the right time. Because 'Everything Works Better When Everything Works Together.'"[64]

Standardization, utility, efficiency, interchangeability. How to destroy the world is as easy as A, B, C.

Oh, we should probably mention that Sybase's big investor is the firm Winston Partners; one of the partners is presidential brother Marvin Bush. As long as we're talking about the use of public office to increase one's private fortune, we should also mention that Sybase has contracts with the U.S. Departments of Agriculture, Defense, Commerce, Treasury, and the General Services Administration; federal procurement database records show 2001 total awards for Sybase of more than $14 million. Winston Partners also have interests in Amsec, which held Navy contracts worth more than $37 million in 2001. Another presidential brother, Jeb

Bush, is an investor in the affiliated Winston Capital Fund, which held U.S. military and NASA contracts worth over $2 million in 2001. And the president's father, Bush I, is of course a principal in The Carlyle Group, which holds devil knows how many contracts and sweetheart deals.[65]

Several associates of former Pennsylvania Governor Tom Ridge now lobby Ridge's new agency, the U.S. Department of Homeland Security, for contracts worth billions.[66] We can rest assured that the granting of these contracts will represent no conflict of interest, because (1) the lobbyists have waited the required interval of one year between leaving government "service" and lobbying their former boss; (2) we all know that Ridge would never under any circumstances give his former associates any sort of special attention whatsoever; and (3) former government officials are found in lobbies everywhere, so this isn't an extreme case.

Whew! We bet you thought there was going to be some corruption there.

Now, it is true that in certain corners of the land, we might be able to find some teensy examples of corruption, but we know that the president of the United States will not tolerate unethical behavior (even if his brother Neil was never brought to justice nor even made to pay back the money he gained for causing the more than one-billion-dollar failure of Silverado Savings and Loan during the 1980s). In George Bush's crackdown on corporate fraud since the Enron inevitability, the Justice [*sic*] Department has achieved "more than 250 corporate fraud convictions, charged 354 people with corporate crime and obtained fines, forfeiture and restitution worth more than $85 million."[67] This may seem like a lot, until you realize the magnitude and routine nature of corporate crime. The majority of Fortune 500 corporations are convicted felons. The annual estimated cost of street crime runs about $2 billion per year. The annual estimated cost of corporate crime runs somewhere in the area of $200 billion.[68] For a quick glimpse into the real relationship between the Panopticon and security (as opposed to security™ as it is sold to us), compare the number of street criminals put away versus the number of corporate criminals put away, and compare the harm each does to society. For a quick glimpse into the ethics of those pretending to halt corporate corruption—sorry for the redundancy—we need merely mention that Deputy Attorney General Larry Thompson, the leader of Bush's corporate fraud prosecutors, "was briefly the subject of controversy when it emerged last year that he was a director of [Providian] credit card company that paid more than $400 million to settle charges of consumer and securities fraud."[69]

Where else does security™ money go?

The CIA has its own venture-capital organization, In-Q-Tel, which funds research into software that could make sense of digital photos or video clips of millions of cars, street signs, and human faces.[70]

Where else? By fall 2003, the United States was spending $3.9 billion a month on military operations in Iraq—with a third of that going to corporations selling food and housing and services, with the major winners, including New Bridge Strategies and Halliburton, enjoying close ties to the U.S. administration.[71]

And so on. It's the golden rule: the one with the gold makes the rules.

We learn, at this late date, that the machine is gold-plated.

The Noose Tightens

It is intolerable to us that an erroneous thought should
exist anywhere in the world, however secret and
powerless it may be. Even in the instant of death we
cannot permit any deviation. In the old days the
heretic walked to the stake still a heretic, proclaiming
his heresy, exulting in it. Even the victim of the
Russian purges could carry rebellion locked up in his
skull as he walked down the passage waiting for the
bullet. But we make the brain perfect before we blow
it out.

George Orwell

There is nothing, no nonhuman, no human, no place, no action,
no thought that is free from the gaze—or at least attempted
gaze—of the watchers at the center of the Panopticon. Scientists
will not be satisfied until they know everything, until they have
entered and penetrated all the holes and corners of the world,
until they have used technological tools and the mechanical
arts—torture—to get the world to betray her secrets fully. And
even then they will not be satisfied, because their eternal dissatis-
faction has never had anything to do with knowledge or its lack,
nor with the world, but with their own wretched and empty
minds and hearts. Within our culture, based on the dyad of guard
and prisoner, powerful and powerless, watcher and watched,
those entitled to exploit and those to be exploited, there can be no
limit to the attempt to control, the attempt to exploit. Those in
power fear everything, so they must attempt to know everything.
They say that *scientia est potentia,* but in their case *scientia derivare
terror*. Their drive for knowledge derives from terror.

• • • •

Let's talk about ECHELON. ECHELON is computer software that searches for specified keywords in e-mail, fax, and telex messages. The keywords cover military activities, drug trafficking and other crimes, the trade in embargoed goods or dual-use technology (that is, commercial goods that can also be used for military purposes), and economic activities.[1] ECHELON is used on the Platform network built in the 1970s and 1980s to fulfill the post–World War II communications intelligence agreement between the United States and the United Kingdom that divided the world into two spheres of influence: the United Kingdom would take Western Europe and the Middle East; the United States would take everywhere else. Canada, Australia, and New Zealand were also party to the agreement, ensuring the global coverage needed for interception of international satellite communications signals.[2]

The communications to be intercepted aren't military; they are private and commercial.

Foreign intelligence routinely includes gathering facts about the economies and political situations around the globe, but there is evidence that ECHELON's surveillance of commercial communications is in effect industrial espionage benefitting corporations based in the home countries (the United Kingdom and United States). This (and the fact that permanent surveillance would be a violation of the European Court of Human Rights) concerned the European Union, which recently established a European Parliament committee to investigate the ECHELON system.[3]

For technical reasons, ECHELON currently has access to only a limited number of communications transmitted by cable or radio, whereas it can intercept telephone, fax, and data signals sent via satellite quite easily, because it has receivers in the right places: Puerto Rico, England, Australia, New Zealand, Japan, Guam, Hong Kong, Canada, Cyprus, Hawaii, and in several locations in the United States (including Fort Meade, the head-quarters of the National Security Agency).[4] Technology to pick out certain voices from millions of telephone recordings is in development.

That's a lot of telephone conversations that can be intercepted, including your latest fight with your long-distance girlfriend, the time your cousin and her boyfriend had phone sex, those hours you counseled your best friend after the breakup of that really shitty relationship he was in a few years ago, your plans to blow up the Grand Coulee Dam, and the time you held up the phone to your cat's face so your niece could hear her purr.

So we return once again to the problem of bureaucrats every-where: the prisoners are having so many conversations, and the guards have so few brains to make sense of them.

That's where ECHELON comes in. Because those in power (neither they, nor their army of agents) can listen in on every com-munication, much as they would like to (especially the one between your cousin and her boyfriend), ECHELON uses "sense-making" software to scan millions of messages for certain keywords.

For now, I guess you're still safe if you do not speak of blowing up the Grand Coulee Dam but instead say, "The grand birthday cake is ready, and the candles are about to be lit. As soon as they are lit, will you be ready to blow them out? It will all be very grand, and cool, eh?"

And I'll give you another hint. If you're going to have phone sex, make sure your partner isn't in another country: all interna-tional phone calls originating or arriving in the United States are routinely recorded.

Oh dear.

But if you really want to have phone sex, or talk about blowing up dams, maybe your privacy isn't in as much danger as the spooks wish—at least not in Germany. About ten million international communications go through Germany every day. About eight hundred thousand of them are sent via satellite. The slowpokes at the German Foreign Intelligence Service are technically capable of putting less than 10 percent of those through a search engine.[5]

How will those in power maintain their control if the Germans are only scanning seventy-five thousand communications per day?

Let's talk about Carnivore. Carnivore is a packet analyzer or "sniffer" box attached to Internet service providers' computers to record data on e-mail traffic. Let's say you're a suspect. If you checked this book out of a library using your own library card or bought it from a bookstore using your credit card, you might be. Carnivore can record the names of the sender and recipients of all your e-mail messages, as well as the messages' contents. It can compile a history of the Web pages you visit. I'll bet that's the last time you visit www.spankmetwice.com.

The FBI explains why it needs Carnivore: "The Nation's communications networks are routinely used in the commission of serious criminal activities, including espionage. Organized crime groups and drug trafficking organizations rely heavily upon telecommunications to plan and execute their criminal activities. The ability of law enforcement agencies to conduct lawful electronic surveillance of the communications of its criminal subjects represents one of the most important capabilities for acquiring evidence to prevent serious criminal behavior."[6]

Just for the heck of it, let's change a few words and see how this FBI statement reads: "The Nation's communications networks are routinely used in the commission of serious criminal activities, including espionage and the activities of large corporations. Organized crime groups, including government agencies, drug trafficking organizations, and corporations, rely heavily upon telecommunications to plan and execute their criminal activities. The ability of law enforcement agencies to conduct lawful electronic surveillance of the communications of its criminal subjects represents one of the most important capabilities for acquiring evidence to prevent serious corporate and other criminal behavior." Can you imagine how different things would be if the FBI actually did investigate corporate criminals?

Back to the one-way world of the Panopticon.

From the perspective of the feds, it's not that big a deal; employers have long been recording what employees send and receive and where they go on the Internet. Why shouldn't law enforcement agencies be able to monitor criminals' and terrorists' e-mail, just like they tap telephones and open mail? As the FBI notes, "Electronic surveillance has been extremely effective in securing the conviction of more than 25,600 dangerous felons over the past thirteen years. [We can wonder how few of those convictions involved those who run major corporations, even though dangerous products kill twenty-eight thousand Americans per year, exposure to dangerous chemicals and other unsafe working conditions in the workplace kills another one hundred thousand, and workplace carcinogens cause 28–33 percent of all cancer deaths in this country.[7]] In many cases there is no substitute for electronic surveillance, as the evidence cannot be obtained through other traditional investigative techniques." In the old days, pen register devices could be used to record telephone numbers that were dialed from the suspect's telephone, and a trap and trace device on a telephone line could determine where a telephone call originated from. But all the FBI got was phone numbers, not the substance of what was being communicated.

Carnivore catapults the FBI from old-fashioned eavesdropper to the ultimate witness. There's no longer any debate about what a terrorist (or foreigner, or activist, or someone who's just plain fed up with the corporate oligarchy that pretends to be a democracy) said or didn't say. "Unlike evidence that can be subject to being discredited or impeached through allegations of misunderstanding or bias, electronic surveillance evidence provides jurors an opportunity to determine factual issues based upon a defendant's own words."[8] Facts and reality, not accusations. Hearing, not hearsay. After all, Rule 901 of the Federal Rules of Evidence requires that evidence be authenticated before it can be admitted as evidence in court.

It gets more interesting: the FBI is developing a Magic Lantern virus that allows Carnivore to collect passwords from computers.[9] Why should law enforcement agencies be prevented from accessing a criminal's computer system? Or, for that matter, maybe the computer system of someone who doesn't like the FBI.

And what will happen in a few years, when cable TV, telephone, and Internet service are all integrated? Drug dealers won't be able to buy hemp clothing online without being found out. Child pornographers won't be able to send dirty pictures illegally. Terrorists won't be able to use e-mails to put together their nefarious international plots (unless of course the terrorists are already at the center of the Panopticon). This layer of security is crucial, which is why the USA PATRIOT Act (passed in collective paranoid delusion six weeks after the 911 attacks) made it easier for the government to get wiretapping authority, makes it clear that wiretapping authority applies to the Internet, and allows (actually, *requires*) banks and credit reporting agencies and stockbrokers and cable TV and Internet service companies to cooperate with law enforcement agencies by providing access to information about their customers—without fear of being sued by their customers for violating privacy laws.[10] We feel much more secure now, don't we?

Since terrorism, like capital, doesn't respect national borders, the cooperation has to be international, which is why thirty-three nations in the Council of Europe plus the United States, Canada, Japan, and South Africa signed the international Convention on Cybercrime soon after 911. This treaty requires the member countries to pass laws against hacking, child pornography, and the theft of intellectual property. It requires each nation to give broad search and seizure authority to law enforcement authorities, including the power to force Internet service providers to cooperate with the use of surveillance technologies such as Carnivore. And it requires each nation's law enforcement agencies to cooperate with the agencies of other member nations.[11]

So often debates about surveillance in our culture boil down to arguments about which is more important: privacy or security. The headline on the Privacy and Technology Web page of the American Civil Liberties Union (ACLU), for example, asks "Is the U.S. Turning Into a Surveillance Society?" and follows this rhetorical question by statements to the effect that "Big Brother is no longer a fiction" and "The United States has now reached the point where a total 'surveillance society' has become a realistic possibility." Yet an ACLU spokesperson responds with the underwhelming "Given the capabilities of today's technology, the only thing protecting us from a full-fledged surveillance society are the legal and political institutions we have inherited as Americans."[12]

It seems those at the ACLU think the solution to government surveillance is to appeal to the government. Why don't we just ask those at the center of the Panopticon to turn off the lights in our cells? And while we're at it, why don't we ask them (nicely) to unlock the doors?

The ACLU claims that its report "Bigger Monster, Weaker Chains: The Growth of an American Surveillance Society"[13] "step[s] back from the daily march of stories about new surveillance programs and technologies and survey[s] the bigger picture."[14] Inevitably, since the ACLU is a union of lawyers, its "bigger picture" is made of laws, and its recommended solution is to "build a system of laws that can chain" the surveillance monster.[15] The ACLU's system would be built of privacy laws, the regulation of new technologies, and a revival of the Fourth Amendment to the U.S. Constitution (protection against unreasonable search and seizure of persons, houses, papers, and effects).[16]

"These privacy gains [*sic*]," the paper states, "can be augmented and many threats to privacy can be overcome if citizens band together for reform and enlightened policy. The hope for progress, in sum, lies in the hands of engaged [I'd prefer enraged, actually] citizens who avail themselves of the legal, technological, and political opportunities to act in the marketplace and the political arena."[17]

We are, once again, supposed to appeal to our captors to give us a break. *You can destroy the world,* we say plaintively, *if you just leave us alone in our bedrooms.*

The appeal to laws is absurd. Let's take a keystone law in the efforts to protect people's privacy: the Privacy Act of 1974. Among other things, it forbids U.S. government collection of information on citizens it isn't investigating. Now, there's some logic and protection for you! And who will be investigated? In the words of former U.S. Attorney General Edwin Meese, "If a person is innocent of a crime, then he is not a suspect." (This would be his version of, "So long as you do what we tell you, you have nothing to hide, and nothing to fear.") In any case, according to a 1989 study by OMB Watch, the Privacy Act should apply to twelve U.S. government agencies maintaining 539 record systems containing 3.5 billion records.[18] An agency is exempted from the Privacy Act when the government decides to investigate you. Or if there's a national security issue at stake. Or if the government *buys* the information about you from a corporate source such as ChoicePoint or Axiom. The FBI has fingerprint and other data on tens of millions of Americans. The U.S. Internal Revenue Service (IRS) and the Department of Health and Human Services maintain the name, address, social security number, and quarterly wages of everyone working in the country. Every state government maintains records of names, addresses, descriptions, and photographs of motor vehicle drivers.

The "dilemma" between "privacy" and "security" confronts a false problem with false choices. Since the foundation of the problem isn't the invasion by technological means of a legal right to privacy, law and technology are not the solutions. The problem is centralization of power, and the answer is the fundamental reconfiguring of power relations.

The poet Antonio Machado wrote, "The eye you see isn't an eye because you see it; it's an eye because it sees you."[19]

As those in power know more and more about you, or rather about your consumer/insurance/demographic fragments, you know less and less about them. Since 911, federal agencies have been directed to resist Freedom of Information Act (FOIA) requests to see government documents. The October 12, 2001 "Ashcroft Memorandum" from the U.S. Department of Justice declared that if there were any "sound legal basis" for withholding information from FOI requesters, the agencies should withhold it, and that the Justice Department would back the agencies up. In November 2001, the Bush administration issued an executive order that seals presidential papers indefinitely, and allows former presidents and their descendants to review and reseal records. In March 2002, the White House ordered federal agencies to withhold information for national security reasons even when the FOIA's exemption for national security does not apply. In March 2003, the Bush administration broadened what government information should be classified secret and gave the CIA veto power over interagency decisions.[20] Since then government agencies, from the Environmental Protection Agency to the Federal Energy Regulatory Commission to the U.S. Geological Survey, have been pulling information from public view. The Panopticon gets ever darker in the center.

In the long-term, if there is to be one, humans must return to a sustainable way of life, albeit in a world depleted by the not-brief-enough blowout called civilization. In the short-term, humans face Smart Dust.

Smart Dust is a "self-contained, millimeter-scale sensing and communication platform for a massively distributed sensor network,"[21] a computer board equipped with a sensor, a microcontroller, and a radio. Currently Smart Dust is a third of the size of a credit card; using nanotechnology, it could eventually be the size of a grain of sand. The motes, as they call them, are able to "sense" temperature, humidity, vibration, air pressure, chemicals, and biological substances. You could use them to detect anthrax, say, but also to monitor the vibrations of factory equipment, and warn if the vibration gets to the point of damaging the machines. You could monitor the temperature in buildings. You could spray thousands of motes over a forest fire to monitor temperatures over the whole forest, with motes transmitting their data from one mote to the next back to a central receiving station. Dozens of motes have already been put into trees on Great Duck Island off the coast of Maine to monitor bird habitat.[22]

Intel and the University of Berkeley are developing the Smart Dust motes. Other folks down at the University of California Los Angeles are working in the Center for Embedded Networked Sensing (CENS) to broaden the applications for these networked computer sensors. CENS wants sensors in roads, parking ramps, traffic lights, factories, airports, farms, hospitals, your car, your home, the groundwater, the plankton communities in the oceans. They want to extend the Panopticon everywhere. The head of

CENS shares her dream: "The average person will be reliant upon and affected by these systems, but if we succeed, the systems will be relatively transparent or invisible. However, this vision is the Holy Grail, and there is a lot of work to do before we achieve it." Here is the place to insert the laugh of a mad scientist who knows she's backed by the full financial and police power of the state. She admits there may be some concerns, but she's willing to address them: "Ultimately, we need the participation of social scientists, as well, to explore the social implications of pervasive monitoring."[23]

Note whose participation she invites: social *scientists*. Not citizens, or tribal elders, or any mumbo-jumbo-like spiritual ways of knowing how to deal with these devices. *Social scientists*. Fellow priests.

It's the same old story.

During the U.S. Civil War, an unmanned hot-air balloon was designed to drop explosives. During the Spanish-American War of 1898, spy photographs were taken from kites. Hitler's V-1 "flying bomb" could carry a 2,000-pound warhead for 150 miles, and killed more than 900 civilians in World War II—a modest indication of what was coming. During the Cold War era, the United States used UAVs equipped with spy cameras over Russia, Cuba, Vietnam, and China. In the 1980s Israel used radio-controlled drones to attract enemy fire, thus locating their enemy's weapons. The Pioneer UAV was used by the U.S. military in the early 1990s to give their battleship commanders a bird's eye view of targets in Iraq.[24]

By the mid-1990s, the U.S. Predator drone was used over Bosnia to give generals (who don't like to get too close to battle) a view of the fighting in real time. The Predator was linked to GPS satellites, could fly for forty hours without refueling, and at an altitude of twenty thousand feet was invisible to those on the ground. It was equipped with infrared radar that could detect heat sources at night, and "see" through clouds. It had a laser beam that locked onto targets, giving missile operators on the ground a sure shot.

By the time the United States invaded Afghanistan in 2001, the Predator was cruising at 65,000 feet, had a 450-mile range, and was transmitting pictures with infrared cameras via high-definition color television. A ground team in a remote control station operated the unmanned plane by radio or satellite link. By the end of 2002, the U.S. Air Force had more than fifty Predators and was producing two more each month.[25]

Soon after the U.S. bombing of Afghanistan began, the Panopticon—we mean the Pentagon—bought up rights to all satellite photos of Afghanistan, thus making sure that it was the only agency that could actually see what was happening. By law the Pentagon has the right to exercise "shutter control" over civilian satellites launched from the United States, but because such an exercise might have made the banned photos subject to the FOIA, the Pentagon cleverly bought exclusive rights to the photos and secured an agreement from the company, Space Imaging, not to "sell, distribute, share or provide the imagery to any other entity."[26]

The Marines used a forty-pound PackBot to search Afghan caves, "bunker-busting" cruise missiles to penetrate them, and "thermobaric" fuel-air explosives to suck the oxygen out of them—as well as good old-fashioned bombs and the "daisy cutter" bombs developed during the Vietnam War that spew enough napalm to kill everyone (by fire, suffocation, or even straightforward disintegration) within six hundred meters. The plane that drops daisy cutters has to fly at six thousand feet to escape the blast.[27] The Marines hope soon to be using a twenty-pound dune buggy called the Dragon Runner, which will be able to cruise streets and enter buildings in urban combat; they already use laptop computers to receive video and sound from their Dragon Eye UAV.[28]

Just as the Panopticon was designed to spy for the purpose of punishment, it's inevitable that these gizmos will be used for more than looking. The Predator was designed for spying but has also been equipped with Hellfire missiles, and the Pentagon and the CIA have used them in Afghanistan, Yemen, and elsewhere to kill the machine's enemies.[29] Hellfire-equipped Predators are flying over Iraq, and the United States is "quietly examining the feasibility of assassinating Hizbullah leaders linked to the deaths of U.S. soldiers and civilians in the 1980s."[30] These are UAVs, some flying without remote control, that can launch missiles at targets selected by computer. Abstract, inhuman, unaware, non-negotiable, fast, silent, lethal: the panoptic ideal married to the aggressor's demands. But there's always room for improvement, if only more efficiency: "Predator B, designated MQ-9B by the U.S. Air Force and referred to as the Hunter-Killer, flies faster, higher and carries more weapons than the Predator."[31]

The Helios vehicle is expected to be able to fly at fifty thousand to seventy thousand feet for months at a time. DARPA is also working on miniature UAVs for military surveillance, law enforcement, and "civilian rescue" operations. One mini-UAV, the Black Widow, has a six-inch wingspan and weighs two ounces. Maybe we'll be able to buy them by the case at Costco.

If you don't want to wait for the ultralite version of the UAVs, you can still take a gander at the Domestic Control Hover Drones (DCHDs) (note the word *domestic*), which the U.S. government is already mass producing and offering for sale to selected governments for $178,000 each (that's for the stripped-down model, the price the salespeople use to get buyers through the door; extras will easily bring the price up to $350,000, not including destination fees). They're shaped like a doughnut a little over three feet across and weigh about forty-five pounds. They have a motor in the middle.

DCHDs are designed to hover about fifty feet up, but they can go as high as five hundred feet. They can stay airborne for three hours without running out of fuel, and they can travel at fifty miles per hour. They can be controlled either by satellite or by vehicles on the ground. They come in your choice of an elegant black matte for night use and a matte white and sky-blue for those days you're feeling a bit more festive. They are, of course, nearly silent, so quiet they cannot be heard from ten feet away.

Did we mention the word *domestic* in the title of the drone?

In case you do somehow see or hear the thing, and want to get away from the state-of-the-art video cameras, you might consider going inside. But that won't help. The drone also has a thermal imaging sensor camera, allowing its controllers to find you through walls.

It also has microphones (state-of-the-art, of course, capable of picking out one conversation among many at more than a quarter-mile), and transmitters so that the controllers can speak to you directly. They can tell you to stop, they can tell you to approach the machine. They can ask you to place your identification card in front of the video camera.

And if you refuse, this Domestic Control Hover Drone is also equipped with a stun-gun.[32]

Did we mention the word *domestic* in the title of the drone?

Technology is not neutral. It does not serve communities. Despite predictions that all these panoptic gizmos will soon be commercially available to you the "consumer," the truth is that "they" (government and corporations) have the resources and capacity to collect and analyze information to control people (consumers, citizens, human beings), and you don't. You're not going to buy a predator drone to eliminate your bothersome neighbor or the oppressive police. You're not going to be privy to the data gleaned from the RFID tags on your neighbors' clothing.

Of course technophiles and technoholics alike will trot out their obligatory arguments that technology is neutral, and that the effects of technology just depend on who uses it, implying that things would be different if only they and not George Bush (insert bin Laden, Saddam Hussein, the Democrats, whoever) controlled predator drones. It's a stupid argument, as pointless in its own way as discussing whether some theoretical Christianity could possibly not commit genocide, some theoretical capitalism could possibly not consume the planet, some theoretical science could possibly not have as its primary goal the attempted control of everything. High technology, Christianity, capitalism, science: these all spring from the same mindset. So *of course* they will all move inexorably toward the same ends, with any pesky legal and moral objections only harrassing their ankles like fleas.

It's a moot point anyway, since the purpose of technology in our culture is to leverage power, and so it is inevitably driven by the ultimate in leverage, the military.

Computers and the Internet were first designed by the military. Many pesticides were originally designed as chemical weapons against humans. Half the scientists and engineers are engaged in military-related research. Many of the technologies used for surveillance, tracking, detaining, and destroying were developed or funded by the military, or quickly adapted to police and military uses.

The director of MIT's computer science lab credited military-supported work with "half of the major innovations in computing, including breakthroughs in microcircuits and data-management systems."[33] This relationship between science and war continues in the marriage of machines and living beings. The U.S. Army has established an Institute for Collaborative Biotechnologies at the

University of California, Santa Barbara, the California Institute of Technology, and MIT. The institute wants to find "better materials for uniforms or armor, faster and lighter computers and batteries and more elaborate sensors."[34]

Then there's the Biometric Consortium, which "serves as the federal government's focal point for research, development, test, evaluation, and application of biometric-based personal identification and verification technology."[35] By 2002, the Consortium had more than eight hundred members, making it sound like some grand community effort. But guess what? The Consortium was initiated and operates under the National Security Agency. The members are government agencies or organizations; those from "private industry and academia will be invited to the Consortium meetings in an observer capacity."[36]

Of course biometrics isn't just for military use. It has many civilian applications. Workplace and airport security. Fingerprinting in Stockholm public schools. The Los Angeles City Hall and the New York City Police Department.[37]

The Consortium isn't alone in promoting the measurement of life. The Biometric Interoperability, Performance and Assurance Working (BIPAW) Group supports "advancement of technically efficient and compatible biometric technology solutions on a national and international basis." The BIPAW Group consists of over ninety organizations "representing biometric vendors, system developers, information assurance organizations, commercial end users, universities, government agencies, national labs and industry organizations."[38] The Group's host is the National Institute of Standards and Technology (NIST), an agency of the U.S. Department of Commerce's Technology Administration. NIST was founded in 1901 with a mission to "develop and promote measurement, standards, and technology to enhance productivity, facilitate trade, and improve the quality of life." Standardization, utility, efficiency, interchangeability.

How to destroy the world is as easy as A, B, C. This obscure little agency spends $864 million a year employing three thousand scientists, engineers, technicians, and support and administrative personnel, plus sixteen hundred guest researchers. And NIST partners with two thousand manufacturing specialists.[39]

NIST is pretty upfront about the way it pervades our lives. It even says on its Web site: "Take a tour of your house and find out where the National Institute of Standards and Technology (NIST) has an unseen role."[40] They say this like it's a good thing.

Where does the military fit in? Well, to answer that, let's ask where the BIPAW Group meets. At RAND's Washington office (also known as Pentagon City) in Arlington, Virginia.[41] Is it a tax-spending government agency? Is it a money-making corporation? Is it a military contractor? Who knows?

Controlling physical access to buildings is one of biometric's more popular uses, with one estimate predicting a market of $389 million by 2004.[42] That's just a fraction of the total market for biometrics, predicted to grow from $116 million in 2000, to $2 billion in 2006, to $5 billion in 2010.[43] After all, if you can sell biometrics to measure schoolchildren, you can push it in the workplace too. Currently less than 1 percent of North American companies use biometrics to secure their computer systems.[44] Bentham's ghost would be proud to know that number is soon to rise.

The U.S. military asks, "What do biometrics do for me and how can I use them? Biometric recognition can be used in identification mode, where the biometric system identifies a person out of the entire enrolled population by searching the database for a match. A system also can be used in verification mode, where the biometric system authenticates a person's claimed identity from his/her previously enrolled pattern. Using biometrics for identifying and authenticating human beings offers some unique advantages. Only biometric authentication bases identification on an intrinsic part of a human being. Tokens, such as smart cards, magnetic stripe cards, physical keys, and so forth, can be lost, stolen, duplicated, or left at home. Passwords can be forgotten, shared, or observed."[45]

Funny, even if the technology were available to those of us in the outer ring of the Panopticon, we don't see how using high technology to "authenticate a person's claimed identity from his/her previously enrolled pattern" will do anything for *us*. We already have our own technologies for doing this: they're called eyes, ears, memory, discernment, and a process called "getting to know a person." We can see, however, how mechanical (and thus far less sophisticated) technologies would be useful for those running a massive prison.

Further, note the sleight of mind that takes place in this sentence: "Only biometric authentication bases identification on an intrinsic part of a human being." They have continued the redefinition and reduction of the human to that which can be measured, sorted. And the statement is just plain false. I know who George is not because of his biometric measurements, but because of *who he is*. And certainly I would say that *who George is* includes what is intrinsic to his humanity. And what is intrinsic to one's humanity? Fingerprints? DNA? Sure. How about need for a community? Meaning? Love? A living landbase? Their definition has eliminated these from intrinsic humanity.

And then there's brain fingerprinting. As usual, it began as an application for the "criminal justice" system, once it was admitted as evidence in court. Now the inventor wants his method to be used in the "fight against terrorism,"™ in the pharmaceutical industry, to measure the effectiveness of advertising, to root out insurance fraud, and as a tool for security screening.[46] It's security gone commercial.

It's scientific of course. It allows us—or rather them—to "measure scientifically" whether a person is a terrorist, spy, criminal, and so on, by finding specific information stored in a person's brain. It can reveal "terrorist training and associations." It can identify "trained terrorists with the potential to commit future terrorist acts, even if they are in a 'sleeper' cell and have not been active for years." It can "identify people who have knowledge or training in banking, finance or communications and who are associated with terrorist teams and acts." It can "determine if an individual is in a leadership role within a terrorist organization."[47] This might even include the capacity to determine whether an individual is in a leadership role within a corporation. It sure would be nice to get some of those high-class crooks with training in banking, finance, or communications off the streets.

As in a lie-detector test, in brain fingerprinting a series of statements are made to a suspect, whose brain responds to familiar information. When the statement is something that only a criminal or the cops should know (since they are part of the same events), then a brain response "implies participation in the crime."[48]

That's all there is to it. How can you argue with that logic? There's no human judgment needed! It's Science! Why? Because "the entire Brain Fingerprinting system is under computer control, including presentation of the stimuli, recording of electrical brain activity, a mathematical data analysis algorithm that compares the responses to the three types of stimuli and produces a determination of 'information present' or 'information absent,' and a statistical confidence level for this determination. At no time during the analysis do biases and interpretations of a system expert affect the presentation or the results of the stimulus presentation."[49] If there's no human emotion or judgment involved, it's *got* to be true. That's one of the rules. And it's been confirmed by the distinguished senator from Iowa: "It seems to me that if we are interested in making sure that the innocent go free, and that the guilty are punished, any technological instrument that can help us make a determination of guilt or innocence, we ought to know about it."[50]

And you say we're not already living inside of the machine?

The End

> We seek alien victims rather than the real source of
> our rage, since we cannot recognize the truth and are
> unable to admit our own despair. And the more we
> express our rage, the more intense it becomes. If we
> don't discover its cause, our pathological state
> worsens, and eventually produces a magical image of
> self and world. We feel invulnerable when we can
> victimize others, even torture them, without recog-
> nizing that it is our own helplessness that is being
> crushed.
>
> *Arno Gruen*

We have surrendered control over our lives—and over our survival, and the survival of most of the planet—to the machines created by our mechanical way of seeing the world. We have
throughout this book been repeating the mantra of the slave, of
the cog: *So long as we do what they tell us, we have nothing to hide,
and nothing to fear.*

But that's all wrong. The truth, in this cult(ure) of the machine,
this cult(ure) of death, is that if you are a living being, you have at
this point very much to fear.

Information and technology are inherently neutral, right? So
how can we say they're the glue that holds science, industry, and
war together? To answer that, just look for an example at the
career of the Assistant Secretary of Defense for Networks and
Information Integration, John Stenbit. He is the War (oops,
Defense) Department's Chief Information Officer, and his recent
title was Assistant Secretary of Defense for Command, Control,

Communications, and Intelligence. For over thirty years he has worked in both public and corporate sectors in the telecommunications and the command and control fields, from Aerospace Corporation (missiles) to TRW (satellite surveillance) to the Science and Technology Advisory Panel serving the U.S. Central Intelligence Agency, to the Federal Aviation Administration's research and development committee, and the National Research Council Manufacturing Board.[1]

Is it war? Is it science? Is it civil engineering? Is it a consumer product? Yes!

Or let's talk about the Pentagon's DARPA. This is the DARPA that's been in the news recently because the head of its Total Information Awareness Office, gunrunnner and spy extraordinaire, John Poindexter, had to resign when the guffaws and gasps over his FutureMap gambling project got too loud. The DARPA that declares *Scientia est potentia*. The Pentagon's Advanced Research Projects Agency (ARPA) was established in 1958, a few months after the Russians sent Sputnik into space. ARPA's first head was General Electric vice president Roy Johnson. Academic and industry researchers have worked together on rockets, ballistic missiles, space travel, nuclear testing, integrated circuits, computer software and networks, grid computing, telepathy and psychokinesis, sensors and radars, fuel cells, navigation and guidance systems, video surveillance, bioinformatics, and radar mapping.[2] In 2003, DARPA will spend $2.7 billion on two hundred projects ranging from computing to space weapons, counterterrorism, drone aircraft, and biological defense.

A few of the many benefits of DARPA research include the M-16 assault rifle, the original Internet for military and government communications, several generations of unmanned drones, the B-2 Stealth bomber, the TIA program, a camera on a chip, the FutureMap online futures market for betting on terrorist events, and a Los Angeles to Las Vegas robot race.[3] DARPA has also developed the Limb Trauma Management Training Simulator to provide a realistic alternative to animal-based and cadaver-based training in modeling gunshot wounds.[4] Part of this effort to understand what happens when they shoot someone is a Bio-Computation Program that's "aimed [naturally] at exploring and developing computational methods and models at the bio-molecular and cellular levels." These methods and models will provide "analytical tools for prediction and control [naturally] of cellular internal processes and systems [*sic*] of living cells, for application in a variety of contexts of interest to DOD." What the heck does prediction and control mean in this case? The "rapid prediction of the impact of external agents and environmental factors, and quick identification of targets and design of intervention mechanisms." We think that means how fast people will fall apart when they've been hit by a piece of metal. But we're not sure.

Anyway, DARPA is looking for proposals from interdisciplinary teams of computer scientists, biochemists, chemists, engineers, and researchers from other relevant fields. They want to look at the "mechanisms [it's all mechanical, all a machine] related to pathogenic processes, their prediction and control; mechanisms that enable discovering powerful computing methods that mimic natural in-vivo computation; [and] mechanisms such as circadian rhythms, the control over which may lead to war fighter effectiveness and well-being in stressed conditions."[5]

Lots of people are saying that "things are getting tougher in the post-911 world," as if things haven't been getting tougher since the advent of centralized power several thousand years ago. Scholars like Mumford and popular culturists like the Wachowski brothers have traced the rise of the machine in fact and fiction.

The police look more and more like soldiers every day—and fashionable women aren't far behind.[6] There's money in it for the

clothing designers, and there's money in it for the police and secu-
rity™ forces, who smell a market. Since 911, local police forces
have been encouraged to see themselves as the first line of defense
against attacks from abroad, and crimes are increasingly investigated as possible terrorist attacks.[7]

But hasn't crime always been confused with war, and citizen-
ship with patriotism? Mustn't those in the outer ring of the
Panopticon always be demonized, called everything from savages
to criminals to terrorists? Isn't terrorists™ simply the latest in a
long line of epithets and excuses for ratcheting up control?
History's heretic is yesterday's communist is today's terrorist.

I look again at the back of a dollar bill. This time I focus not on
the all-seeing eye, but on the truncated pyramid below. I'm
thinking about the culture as a pyramid (and a pyramid scheme)
where those at the top—those whose eyes characterize the
Panopticon, and who are themselves overseen by the beneficent
eye of providence—gain from the work of those below. Lewis
Mumford wrote of this mode of organization: "The social
pyramid established during the Pyramid Age in the Fertile
Crescent continued to be the model for every civilized society,
long after the building of these geometric tombs ceased to be fashionable. At the top stood a minority, swollen by pride and power,
headed by the king and his supporting ministers, nobles, military
leaders, and priests. This minority's main social obligation was to
control the megamachine, in either its wealth-producing or its
illth-producing form. Apart from this, their only burden was the
'duty to consume.' In this respect the oldest rulers were the pro-
totypes of the style-setters and taste-makers of our own over-
mechanized mass society."

Sound familiar? Then try this on, too, "Henceforth, civilized
society was divided roughly into two main classes: a majority con-
demned for life to hard labor, who worked not just for a sufficient

living but to provide a surplus beyond their family or their immediate communal needs, and a 'noble' minority who despised
manual labor in any form, and whose life was devoted to the elaborate 'performance of leisure,' to use Thorstein Veblen's sardonic
characterization. Part of the surplus went, to be just, to the support of public works that benefited all sections of the community;
but far too large a share took the form of private display, luxu-
rious material goods, and the ostentatious command of a large
army of servants and retainers, concubines and mistresses. But in
most societies, perhaps the greatest portion of the surplus was
drawn into the feeding, weaponing, and over-all operation of the
military megamachine."[8]

Stealth bombers, anyone?

Whether we're talking about dysfunctional abusive families or
dysfunctional abusive cultures, social pyramids are never based
(at least not for very long) exclusively on naked force (although
force always underlies them). A basis in naked force is not
pleasing to those in power, who cannot allow themselves to see
the evil they have become, nor to those without power, who long
for a figleaf of dignity, if they can't have freedom. Instead the
atrocities must be fully rationalized by all. Kings (and modern
presidents) rule not just because they've convinced a bunch of
armed men (commonly called soldiers or police) to kill for them,
but because God says they have the divine right to do so. When
the machine arrived in North America, its servants did not take
the land just because they had more weapons and different
morals than the humans who already lived here, but because their
every mechanical step was assisted by providential agency. Now,
the servants of the machine do not kill the planet because they
have a death urge, and because they have convinced hundreds of
millions of men and women to participate in the corporate
economy—to support production, to support the conversion of
the living to the dead—and because they have convinced millions

more to be ready to kill to support the actions of this economy, but because, as I read just yesterday in an editorial in the *San Francisco Chronicle*, "the system of capitalism and industrialization . . . leads us, properly, to regard nature as only a means to satisfy man's [*sic* and, frankly, sick] wishes."[9]

Lewis Mumford wrote that "the earliest complex power machines were composed, not of wood or metal, but of perishable human parts, each having a specialized function in a larger mechanism under centralized human control. The vast army of priests, scientists, engineers, architects, foremen, and day laborers, some hundred thousand strong, who built the Great Pyramid, formed the first complex machine, invented when technology itself had produced only a few simple 'machines' like the inclined plane and the sled, and had not yet invented wheeled vehicles."[10]

In his monumental *Myth of the Machine,* Mumford made explicit the relationship between mechanization and control: "Conceptually the instruments of mechanization five thousand years ago were already detached from other human functions and purposes than the constant increase of order, power, predictability, and above all, control. With this proto-scientific ideology went a corresponding regimentation and degradation of once-autonomous human activities: 'mass culture' and 'mass control' made their first appearance." Mumford's words ring true today.

It really isn't possible to think about pyramids without remembering their purpose. Similarly it really isn't possible to talk about mechanization without talking about our culture's obsession with death. Mumford brings them together: "With mordant symbolism, the ultimate products of the megamachine in Egypt were colossal tombs, inhabited by mummified corpses; while later in Assyria, as repeatedly in every other expanding empire, the chief testimony to its technical efficiency was a waste of destroyed villages and cities, and poisoned soils: the prototype of similar 'civilized' atrocities today. As for the great Egyptian pyramids, what are they but the precise static equivalents of our own space rockets? Both devices for securing, at an extravagant cost, a passage to Heaven for the favored few."[11]

Death isn't an incidental by-product of the machine. The obsession with security ends in wealth-making as ecocide and war as genocide. The obsession with security becomes a death wish not because of the outward appearance or the inner character of the perpetrators; it's an obsession rooted in underlying, often unconscious motivations. Sometimes it manifests as bloodthirsty, sometimes as larcenous. Sometimes as an overzealous concern for the well-being and right living of others. Sometimes as patriotic pride or dedicated work ethic. Sometimes as pathological lying, as mindless vandalism, or as a good citizen's respect for the rule of law. But if the underlying motivation is security and rules over freedom, then the end result, the collective result, is to choose comfort over tolerance, rules over responsibility, known over uncertain, wealth over wilderness, control over relationship, and ultimately, death over life. Life isn't a simplistic set of opposites with one obvious right answer to every question, but the obsession with security tries to make it so.

The smoke and mirrors of high technology appear to be, Wizard-of-Oz-like, our masters. But the fearsome wizard behind the curtain turns out to be everyman and everywoman, fussing with levers, no longer in control or quite sure what the purpose of it all is. We've become caretakers of the machine, janitors in the machine's warehouse, sweeping and tidying up and oiling the levers and attaching the new gizmos that are supposed to control this or that section of the machine. (How come those gizmos are called *governors*?) We aren't sure anymore what it's all for, but someone somewhere must be keeping track, we tell ourselves, and those engineers sure are clever. And we really can't imagine how the machine could be replaced with anything else. It's big and it's more complicated than anyone can imagine, and it's all humming on into the future. . . .

Gee-whiz science is transforming our world. Isn't it amazing, this technology that can be used for good and ill? One of the world's "most advanced humanoid robots" is feted for "walking, turning and even dancing with children."[12] But it's not just for kids! When the Japanese prime minister paid an official visit to the Czech prime minister in the summer of 2003, he brought along a four-foot-tall robot that could tell jokes and make a toast. The robot offered to dance, but the Czech prime minister is not into dancing, and declined.[13] Dancing isn't the only thing robots are good for! The Pentagon's DARPA is offering a million bucks to see if someone can create a machine capable of going from Los Angeles to Las Vegas without human intervention.[14] The purpose of DARPA's "autonomous ground vehicle" race in March 2004 was to "leverage American ingenuity to accelerate the develop-ment of autonomous vehicle technologies that can be applied to military requirements."[15]

Big-tech apologist Witold Rybcynski begins his book *Taming the Tiger* with the declaration that "we must live with the machine; we have little choice"—marshaling what has become, pathetically enough, one of the primary arguments in favor of the current deathly system: we're committed to it, so if you don't like the death of the planet, tough luck—and ends it with the sugges-tion that we can control technology "by directing its evolution, by choosing whether and how to use it, or by deciding what signifi-cance it should have in our lives." But he conveniently forgets that civilization is predicated on power differentials. How do I direct the evolution of the unmanned predator drone? How do we choose whether to use the forty million closed circuit television cameras that capture our movements down the street? How do I decide what significance the recording of my e-mail messages and the revoking of my biometric-encoded passport shall have? Rybcynski rightly reminds us that "the struggle to control technology has all along been a struggle to control ourselves."[16] We say rightly, but we mean partly rightly, since we aren't the ones creating predator drones. We aren't the ones at the center of the Panopticon. This is why it's not a matter of controlling technology, but of changing the power relations in society. Our obsession with comfort makes us addicts to technology, and our attachment to security makes us servants of authority. As addicts and servants we neither control technology nor change the nature of power.

Jean-Jacques Rousseau's *Social Contract* declares that governance is by consent, not by mandate. He wrote, "Man is born free; and everywhere he is in chains." He also wrote, "The brain may become paralyzed and the individual still live. A man may remain an imbecile and live. . . ."[17] We see both of these around us each day. And he wrote further, as we also see each day, "Slaves lose everything in their chains, even the desire of escaping them. . . . If then there are slaves by nature, it is because there have been slaves against nature. Force made the first slaves, and their cowardice perpetuated the condition."[18]

Rousseau was perhaps following the analysis of Éttiene de La Boétie, who in 1564 wrote, "Poor, wretched, and stupid peoples, nations determined on your own misfortune and blind to your own good! You let yourselves be deprived before your own eyes of the best part of your revenues; your fields are plundered, your homes robbed, your family heirlooms taken away. You live in such a way that you cannot claim a single thing as your own; and it would seem that you consider yourselves lucky to be loaned your property, your families, and your very lives. All this havoc, this misfortune, this ruin, descends upon you not from alien foes, but from the one enemy whom you yourselves render as powerful as he is, for whom you go bravely to war, for whose 'greatness' you do not refuse to offer your own bodies unto death. He who thus domineers over you has only two eyes, only two hands, only one body, no more than is possessed by the least man among the infinite numbers dwelling in your cities; he has indeed nothing more than the power that you confer upon him to destroy you. Where has he acquired enough eyes to spy upon you, if you do not provide them yourselves? How can he have so many arms to beat you with, if he does not borrow them from you? The feet that trample down your cities, where does he get them if they are not your own? How does he have any power over you except through you? How would he dare assail you if he had no cooperation from you? What could he do to you if you yourselves did not connive with the thief who plunders you, if you were not accomplices of the murderer who kills you, if you were not traitors to yourselves? You sow your crops in order that he may ravage them, you install and furnish your homes to give him goods to pillage; you rear your daughters that he may gratify his lust; you bring up your children in order that he may confer upon them the greatest 'privilege' he knows—to be led into his battles, to be delivered to butchery, to be made the servants of his greed and the instruments of his vengeance; you yield your bodies unto hard labor in order that he may indulge in his delights and wallow in his filthy pleasures; you weaken yourselves in order to make him the stronger and the mightier to hold you in check. From all these indignities, such as the very beasts of the field would not endure, you can deliver yourselves if you try, not by taking action, but merely by willing to be free. Resolve to serve no more, and you are at once freed. I do not ask that you place hands upon the tyrant to topple him over, but simply that you support him no longer; then you will behold him, like a great Colossus whose pedestal has been pulled away, fall of his own weight and break into pieces."

He also wrote that it is "the inhabitants themselves who

permit, or, rather, bring about, their own subjection, since by ceasing to submit they would put an end to their servitude. A people enslaves itself, cuts its own throat, when, having a choice between being vassals and being free men, it deserts its liberties and takes on the yoke, gives consent to its own misery, or, rather, apparently welcomes it."[19]

A few of the layers of modern society maintain some semblance of consent, but amidst all the high-tech communication and the waves of data, is it informed consent? Is it the consent of consumers who purchase the latest piece of technology wrapped in layers of disposable ancient forest paper? Is it the consent of employees who are utterly dependent upon their paychecks for food, clothing, and shelter? Is it the consent of the wards of high-tech hospitals, unprepared to die, who consent to millions of dollars worth of invasive medicine to prolong life for another month? Is it the consent of voters who choose between corporate-funded and party-chosen candidates?

In this book we have discussed many examples of brittle technologies, the secret machinations of politicians and executives, and the underlying sorting of society by anonymous bureaucrats. The more we depend on these systems, the more insecure we feel, and the more we feel beholden to the latest technology, the latest demagoguery, the latest parental projection for safety. Freedom is responsibility. Security is slavery. Denial doesn't change this.

The world's largest administrative building is the Pentagon; the second-largest is the former residence of the Romanian dictator Ceausescu.[20] Make of that what you will.

The technophiles don't necessarily believe they are security-driven and fear-obsessed. They are just citizens trying to get along. As a *Los Angeles Times* business writer put it, "The 'geeks' who once shunned activism amid the digital revolution are using their money and savvy to influence public policy. . . . After years as political agnostics, the programmers and engineers who orchestrated the technological revolution of the 1990s are trying to reboot government. Top technology executives such as Bill Gates found their public voice years ago. Now, the tens of thousands of technology workers who toiled in cubicles writing software and creating gadgets are making their influence felt. They have money, earned during the boom. They have time, found since the bust. And they are using their technological savvy to recruit even casual Internet users to their causes. They want to make sure civil liberties aren't trampled in the push for greater security. They want privacy respected. And they want the media and the political conversation in general to be freed from the dominance of a small number of powerful groups and corporations. Otherwise, they are hard to place on the political spectrum."[21]

Perhaps the author is speaking of a mythical spectrum that defines political relations in some fairy tale of democratic governance. I look around, and I don't see it. I look at the history of our culture, and I still don't see it. The time doesn't exist in our culture where guard and prisoner commingled, or better, never donned their uniforms. Today, the political spectrum is defined by the degree of one's support for governance by and for the corporation, and the degree to which that support is swayed by thoughts about privacy, civil liberties, and other such minor concerns.

Commerce and governance are more than intertwined—they are inseparable.

The main Web page for the UN's World Summit on the Information Society (WSIS) states, "The global information society is evolving at breakneck speed. The accelerating convergence between telecommunications, broadcasting multimedia and information and communication technologies (ICTs) is driving new products and services, as well as ways of conducting business and commerce. At the same time, commercial, social and professional opportunities are exploding as new markets open to competition and foreign investment and participation. The modern world is undergoing a fundamental transformation as the industrial society that marked the 20th century rapidly gives way to the information society of the 21st century. This dynamic process promises a fundamental change in all aspects of our lives, including knowledge dissemination, social interaction, economic and business practices, political engagement, media, education, health, leisure and entertainment. We are indeed in the midst of a revolution, perhaps the greatest that humanity has ever experienced. To benefit the world community, the successful and continued growth of this new dynamic requires global discussion."[22]

But you and I both know who will be doing the discussing, or more to the point, making the decisions.

The WSIS is intended to "provide a unique opportunity for all key stakeholders to assemble at a high-level gathering and to develop a better understanding of this revolution and its impact on the international community." These key stakeholders are explicitly identified as "Heads of State, Executive Heads of United Nations agencies, industry leaders, non-governmental organizations, media representatives and civil society."

Where are we in this process? Do you have a stake in these decisions? Do I? Do the San Bushmen, the Agta, the Karen, the Guarani, the Mapuche, or any other traditional indigenous peoples whose lives are being consumed by technological society? Do nonhumans, who are being driven extinct while those in power fiddle with the controls of the machines that now own us all? None of the (human or nonhuman) people on the outer ring of the Panopticon are considered stakeholders by those on the inside.

The stated purpose of the WSIS is "to develop and foster a

clear statement of political will and a concrete plan of action for achieving the goals of the Information Society, while fully reflecting all the different interests at stake." Not to beat a dead planet, but once again we know what those in power mean and do not mean by the word "all."

The WSIS organizing committee, which operates "under the patronage" of UN Secretary-General Kofi Annan, is composed of numerous agencies of the UN, but also the director general of the World Trade Organization, the executive director of UNITAR, the president of the World Bank, and the secretary-general of the International Telecommunication Union.[23]

Who are the industry leaders participating in the WSIS? They include the Business Council for the UN, the Global Business Dialogue on Electronic Commerce, the Global Information Infrastructure Commission, the Money Matters Institute, the United States Council on International Business, the World Economic Forum, and other industry associations.[24] But really, *who are these guys?* These associations are made up of executives from AT&T, France Telecom, Alcatel, Masreya, the International Chamber of Commerce (ICC), and other heavyweights.[25]

For example, the ICC, which is itself led by executives from Fransabank, United Airlines, Nestle, etc., is a policy-making body that concerns itself with everything from advertising and marketing to arbitration, banking, business in society, business law, commercial crime, commercial practice, competition, customs, economic policy, e-commerce, energy, environment, extortion and bribery, financial services and insurance, intellectual property, taxation, telecommunications, trade and investment, and transport. You may not know much about the ICC, but the ICC is clear about itself; its Web site states that the ICC World Council is "the equivalent of the general assembly of a major intergovernmental organization. The big difference is that the delegates are business executives and not government officials."

The Web site encourages corporations to "look to ICC as they meet the challenges of globalization and adjust to a world in which the state's role in the economy is no longer pre-eminent."[26]

So the agenda and outcome are clear: the spread of technology and global trade as ends in themselves, unfettered by even nominal divisions between commerce and governance, much less by bleeding-heart concerns over the natural world, the carrying capacity of the earth, or the survival of the majority of humans, whose bread and board have been tenuous for the past several centuries.

It is not enough for the police to police us. Far more important is that we police ourselves. In 2002, the U.S. government created a new Terrorism Information and Prevention System (TIPS), which was to begin with a pilot program in August of that year with a million informants in ten cities. That was only to be the first stage. TIPS was set to grow into "a nationwide program giving millions of American truckers, letter carriers, train conductors, ship captains, utility employees, and others a formal way to report suspicious terrorist activity." In response to criticism, the federal government emphasized that TIPS was "first and foremost . . . a program under development, and its blueprint is not yet finalized."[27] It also undoubtedly started dossiers on the critics.

I had lunch with someone who spent time in prison for committing an act of sabotage against a major corporation. Her environmental group had been infiltrated by one FBI undercover agent and one informant. These federal "assets" had worked for a couple of years to gain the trust of those in the group. The informant even began fucking one of the environmentalists (all things considered, it would be inappropriate to say he started a relationship with her, or that he made love with her, or even that he had sex with her: in fact, you could make the argument that using the word *fuck* is generous, since sex under false pretenses is rape). Eventually the two feds suggested a target for sabotage, supplied the materials, and did everything but commit the sabotage themselves. Oh, or get arrested and sent to prison.

I knew all that before lunch. But I learned something new.

The woman I had lunch with said, "You know how the feds get most of their informants, right?"

I was pretty sure I did, but I wanted to hear her say it.

She continued, "They catch someone committing a crime, and then blackmail the person into cooperating. If the person doesn't inform on others, it's off to prison. If the person does inform, it's off to the bank for payday."

Sticks and carrots, I thought. *Carrots and sticks*.

She asked, "Do you know what crime they were using to blackmail this informant?"

"Drug abuse, wasn't it?"

"Oh, he was an abuser, all right, but that isn't what they got him for."

"I never heard anything else."

"That's because it was never allowed to be spoken of in court, nor was it ever allowed to be reported. But someone did mention it once, before the judge ordered it stricken from the official record." She stopped, then continued, "Striking it from the official record is not the same as striking it from my memory. Do you know what he did?"

I shook my head.

"He's a child molester."

"They set a child molester free to try to catch environmentalists?"

"Not catch," she said. "Set up. Those are their priorities. And I'll tell you something else. Many of us in the group they infiltrated had young children. You do the math."

• • • •

In her book *Eichmann in Jerusalem: A Report on the Banality of Evil,* Hannah Arendt observed, "The trouble with Eichmann was precisely that so many were like him, and that the many were neither perverted nor sadistic, that they were, and still are, terribly and terrifyingly normal. From the viewpoint of our legal institutions and of our moral standards of judgment, this normality was much more terrifying than all the atrocities put together. . . ."[28]

Thomas Merton said much the same thing: "One of the most disturbing facts that came out of the Eichmann trial was that a psychiatrist examined him and pronounced him *perfectly sane.* . . . We equate sanity with a sense of justice, with humaneness, with prudence, with the capacity to love and understand other people. We rely on the sane people of the world to preserve it from barbarism [*sic*], madness, destruction.[29] And now it begins to dawn on us that it is precisely the *sane* ones who are the most dangerous."[30]

Here is what modern sociologists say about the sanity of individual Nazis: "By conventional clinical criteria no more than 10 percent of the Nazi SS [the acronym for *Schutzstaffel,* German for *Protective Corps*] could be considered 'abnormal.' This observation fits the general trend of testimony by survivors indicating that in most of the camps, there was usually one, or at most a few, SS men known for their intense outbursts of sadistic cruelty."[31] The sociologists conclude, "Our judgment is that the overwhelming majority of SS men, leaders as well as rank and file, would have easily passed all the psychiatric tests ordinarily given to American army recruits or Kansas City policemen."[32]

How do we define sanity in a culture that is killing the planet? How do we define rationality? How rational is it to remove 90 percent of the large fish from the oceans? How rational is it to remove 90 percent of the native forest from the land? How rational is it to change the climate? How rational is it to put so many pollutants into the air that infants exceed a lifetime's "safe" exposure to carcinogens in two weeks in Los Angeles, three weeks in San Francisco? How rational is it to have carcinogenic chemicals in half the municipal drinking water in the United States? *No problemo,* you say, *I'll just drink from my well.* Yesterday I read that the U.S. Congress is set to relax the rules allowing oil corporations to further pollute groundwater.

Or maybe all of this *is* rational. It all depends on what you want. It was said of someone whose name you may know, "From insane premises to monstrous conclusions, Hitler was relentlessly logical."[33]

We can say the same for much of our culture's rationality. If your premises are flawed, it doesn't matter how rational you pretend to be; your actions will still be absurd.

What's the point?

Humanity

Listen now what the land is telling us—all of us,
remnant orphans of pillaged and scattered tribes. It is
telling us that the time has come, that the empire's
days are running out, that it is time to begin a dance,
a variegated, brilliant Ghost Dance for the end of the
millennium, that will bring together all the broken
villages, the runaway slaves, the rebels who have
regained their humanity by deserting the machine,
all those who are seeing the signs and beginning to
understand that it was always a lie, and that one way
or another it will come to its end.

David Watson

Nanotech, biotech, artificial intelligence, smart cards, embedded
chips, smart dust, fragmentation of humans into consumer,
patient, voter, worker roles for sorting by government and corporation: all of these together spell the end of the human in ways we
can't entirely grasp, much less control. Are we human? Are we
bionic free agents? Are we cogs in the machine?

What do you want?

What do you love?

Do you love?
    Where is love in all this talk of Panopticons and pyramids, smart
dust and artificial intelligence? Where is the human animal?
    Who are you?

· · · ·

We have some questions. They are not meant to be rhetorical. Who owns identity? Who owns information? What does intellectual property mean? How could it be controlled?

> You'd better watch out, you'd better not cry
> You'd better not pout, I'm telling you why
> Santa Claus is coming to town
> He's making a list, he's checkin' it twice
> He's gonna find out if you're naughty or nice
> He sees you when you're sleeping
> He knows if you're awake
> He knows if you've been bad or good
> So you'd better be good for goodness sake.[1]

Bullshit. It's not about being good for goodness sake; it's about whether you're going to get the goods by being a good boy or girl. Or maybe a good employee, consumer, citizen, tenant, prisoner. So long as we do what they tell us, we have nothing to hide, and nothing to fear, and we can get some presents from the big man at the center of the Panopticon. If we're Christians, this means an eternity with the Big Man himself.

Just don't rock the boat.

Within this panoptic way of perceiving the world, which is the world into which I was baptised as a child, and the world into which all of us are trained early on, children are chattel. So are "adults."

Let's take a stroll through the dictionary.

Chattel: Middle English *chatel,* movable property, from Medieval Latin *capitale.* 1. Law: An article of movable personal property. 2. A slave.

Cattle: 1. Any of various chiefly domesticated mammals of the genus Bos, including cows, steers, bulls, and oxen, often raised for meat and dairy products [and the genus Homo, raised for labor, sex, military service, and so on]. 2. Humans, "especially when viewed contemptuously or as a mob." From the Middle English *catel,* property, livestock, from Old North French, from Old Provençal *capdal,* from Medieval Latin *capitle,* holdings, funds, from Latin *capitlis,* principal, original, from *caput,* head.

Bondsman: circa 1713 (in the time of the British enclosure). One who assumes the responsibility of a bond. 1. A slave; a villain; a serf; a bondman. 2. (Law) A surety; one who is bound, or who gives security, for another.

Bond: something that restrains. An obligation made binding by a money forfeit.

Mortgage: Etymology: Middle English *morgage,* from Middle French, from Old French, from *mort* dead (from Latin *mortuus*) + *gage*—more at *murder*.

Murder: Etymology: Middle English *murther,* from Old English *morthor;* partly from Middle English *murdre,* from Old French, of Germanic origin; akin to Old English *morthor;* akin to Old High German *mord,* murder, Latin *mort, mors,* death, *mori,* to die, *mortuus,* dead, Greek *brotos,* mortal. Date: before twelfth century.

So, prior to the twelfth century, mortgage was murder. By the fifteenth century, when enclosure of the commons was underway, the murder in it was obscured, and it had become a legal instrument of property.

As in, "If you don't pay the rent, your caput is kaput."

Kaput: 1. Utterly finished, defeated, or destroyed. 2. Unable to function; useless. 3. Hopelessly outmoded.

As in *property* and *privacy.*

Property and privacy are related. Ownership of identity is (like) ownership of property. Trying to protect privacy (identity) with laws is (like) trying to treat information as property.

Privacy is failing to protect our selves. Intellectual property is failing to protect human culture, human heritage. As legal instruments, they are failing. Law and property aren't the answer, and neither is competition nor the kind of personal "freedom" nor power that undermines the freedom of others. Community relations require both less law and less property. In community, identity as well as goods are a gift.

We know that traditional functioning human communities have from the beginning of human existence been based on the exchanges of gifts, exchanges where relationships are more important than goods. We know that modern culture is organized around markets, where goods are more important than relationships.

We know that many traditional human communities have no privacy. People know each other. The quest for privacy develops as a response to oppressive relationships. Within real communities, where relationships are not oppressive, the central panoptic statement "So long as we do what they tell us, we have nothing to hide, and nothing to fear" can be inverted to "Because we have nothing to fear, we are free to do as they tell us (or not), and we need hide nothing." In fact, there's no "they" telling us what to do, and nothing to hide.

Think about it. What need would you have for privacy if you had no fear of anyone using what you do against you? That's not to say you won't have modesty or secrecy, or things that you keep to yourself for reasons known only to yourself, for you very well may. But there is a world of difference between not sharing from modesty or secrecy or simply because you don't want to, and not sharing from privacy. Privacy is based on fear, and it is based on unequal power in relationships. The need for privacy is a product of living in the Panopticon. The word itself arose in the fifteenth century (think enclosure) to mean "the quality or state of being apart from observation; freedom from unauthorized intrusion."

We know that traditional human communities have been based on social relations, rather than rules. Power was distributed, and fluid, based on circumstance and experience. Within modern culture, power is concentrated, and static. The rule of law is used to try to control what used to be shared in relationships. Power relations of course lead to irresponsibility: if relations are more important than rules, I am responsible to the relationships, but if relationships are not primary and are based on rules and control anyway, I will try to use or avoid those rules, to twist them to my advantage. Responsibility disappears, to be replaced by moral strictures. Relationships disappear, to be replaced by lawyers.

The most dangerous, foolish, and false thing any of us can possibly say is the mantra of, "So long as we do what they tell us, we have nothing to hide, and nothing to fear." The degree to which we internalize the laws and rules of the machine is the degree to which we have no hope of survival, no hope of escape, and certainly no hope of smashing the machine (which means, once again, no hope of survival). The degree to which we internalize the laws and rules of the machine is the degree to which its control is complete, and self-discipline and self-responsibility become moot.

Toward the end of Orwell's *1984,* a secret policeman says to one of his victims, "The command of the old despotisms was 'Thou shalt not.' The command of the totalitarians was 'Thou shalt.' Our command is 'Thou art.' "[2]

The control the powerful wish to exert (and the powerless wish to obey) extends not only over the wilderness, but into the most intimate areas of our minds and hearts. If knowledge is power, as the Information Age cliché has it, and as DARPA contends, then if they know us, they can control us.

Technology separates us from nature, and social indoctrination—our training from birth on, our molding into machine-shape—separates us from ourselves. When an "external" entity (either hard-wired, like the mass media or the government, or soft-wired, like my obsession with reading the news, or obeying Officer Friendly) controls the information I receive, it controls my experience of the world. And because my experience of the world controls my actions, whosoever controls my experience of the world controls me.

Those in power have known that all along. By now, we should know it, too.

We block caller ID. We encrypt our e-mail. Public interest organizations file class action lawsuits to protect our privacy. Individuals who have been discriminated against go to an arbiter or to court to protect their civil rights. We give the supermarket clerk a false address when we sign up for their discount card, so we won't be the target of junkmail. Those of us who like to drive fast install radar detection devices in our cars. We refuse cookies offered to us by Web sites. We get paid under the table to avoid taxes. We lie for a higher purpose or to make ends meet. We use the law and counter-technologies to protect ourselves from the prying eyes of the market and the governors.

But legal actions and technical fixes only go so far. They don't stop the insinuation of the market into our homes and our minds, and they don't stop the tightening of the noose of government around our necks.

Resistance takes as many forms as the offenses it opposes. The more unequal the power between the machine and the human beings, the more varied and sometimes devious will be the resistance. According to political scientist James C. Scott, the "everyday resistance" adopted by the seemingly powerless includes, among many other forms, "foot dragging, dissimulation, false compliance, pilfering, feigned ignorance, slander, arson, [and] sabotage" as well as poaching, squatting, desertion, and evasion.[3]

Massive enough resistance sometimes changes a policy or stymies a particular program, for a while. For example, in the 1970s, so many people refused to participate in a census in the Netherlands that it became unworkable.[4] If folks get clear and angry enough, they may switch from underground resistance to taking collective action, including petitions, demonstrations, boycotts, strikes, revolt, and revolution.[5] If we're really radical we rail against technology itself, maybe rip a few genetically tortured strawberries out of fields, or torch bulldozers on the edges of ancient forests. Feels good, but doesn't particularly slow the machine. The nature of the system doesn't change. When rebellion goes aboveground, the machines crank up their panoptic sorting of resisters. In times of real unrest, of revolt or war, the governors' only concern becomes the continuation of the state that they rule and enjoy (or at least perceive they enjoy), the continuation of the machine they serve. The system kicks into high gear to identify and sort the acceptable cheaters from the felonious thieves from the labor organizers and public interest troublemakers from the revolutionaries who need to be eliminated.

When I taught at the prison, several of my students commented that judges know how to deal with people who steal because of greed, probably because they know that motivation so well themselves. But people who steal because they hate the system and because they want to bring it down confuse and scare judges, who respond by handing out sentences far beyond the norm. That is the panoptic sort in action, and it does not require particularly conscious cruelty on the part of cops and judges and passive bystanders.

"Some types of resisters—like the upper-middle-class tax shirker—are tolerated, even smiled upon, by political leaders. Others, like the poor women [in welfare programs], are vilified and hunted."[6] The worst of them are labeled terrorists (thieves, witches, gypsies, foreigners, communists, terrorists, the brand of the century) and burned at the stake or shot dead on sight. The liberals cluck their wine-soaked tongues, and ask the lawyers to check on the state of their civil rights. Yep, looks like there are laws against discrimination all right. Somebody ought to enforce them!

How far do we want to go? How free do we want to be? Do we want the machine's widgets and baubles at low prices, or do we want to drop our chains of iron and gold and cybermetal and walk free?

Marx (like most labor unionists, and most environmentalists, and most social justice activists) didn't protest the use of technology or the enjoyment of its benefits; if the benefits weren't being enjoyed by everyone, then it was the masters, and not the workers, who were to blame. But at this late date, let's do more than take our masters to task. Let's task ourselves as well, and admit our role in maintaining the machine. And let's take into account the many externalities crucial to its perpetuation. The masters' pool of unemployed keeps our wages low. Several wild animals are killed for every one who arrives at the zoo, and most of the ones who do arrive alive are dead within a year or two. Automobiles and furnaces waste most of the potential energy in the fuel they burn. The candy bar that takes thirty seconds to eat comes wrapped in an insidious plastic-aluminum amalgam that will flutter about the landscape decades after the candyeater is dead of diabetes. This future trash isn't virtual. It will exist physically, in an actual place, leaching actual toxins into actual rivers and bloodstreams. The candy trashes our blood sugar, but that's nothing compared to what the wrapper will do to our great-grandchildren's blood. Could the shame we feel about our uncontrollable sweet tooth actually be misplaced horror at the devastation caused by every clank of the panoptic vending machine that we trigger with our obsession for cheap sweetness?

The social critic Klaus Lenk wrote, "The real issue at stake is not personal privacy, which is an ill-defined concept, greatly varying according to the cultural context. It is the power gains of bureaucracies, both private and public, at the expense of individuals and of the non-organized sectors of society, by means of the gathering of information through direct observation and by means of intensive record-keeping."[7]

Even more real is the social organization we have been calling the machine versus the life on this planet. The real issue is whether we should tolerate—or even if we can survive—a social organization (including technologies, because technologies emerge from specific social organizations) based on the machine, that is, one that converts all fuels—all life on the planet, including our own—into hierarchically organized power. We say the real issue at stake is how we can stop the machine.

For those who say it's not so bad, look at all the comfort, all we can say is include the poor (that is, the majority of humans) and the nonhuman in your calculations. We can't all beat the system and retire rich in Palm Beach. And what's the use of retiring rich on a dying planet? Those of us who aren't buying into the machine can't beat the machine on its own terms either, without getting thugged. The people at the ACLU and the public defender's office and places like that are really smart, and they're doing their best, but they seem to be getting nowhere fast trying to use law and technology to repair law and technology.

Shall we continue to lull ourselves into sappy enjoyment of the sticky virtual fruits of the machine? Shall we continue to resist but end up alternating between devotion to "simple solutions" and collapsing in a heap of "radical despair?"[8]

Because it converts everything into hierarchically organized power, and because its deadly efficiency causes it to be able to "outcompete"—code language for destroy—the cultures and beings who live in reciprocal relations with their surroundings, it is difficult to take on the machine on its own terms. John Henry blew out an artery trying to outperform the machine. The

Luddites smashed a few machines before those at the center of the Panopticon hung them. Identify with the machine and you will die. Try to destroy the machine and those in power will try to kill you.

What do we do?

How's this for a first step? Let's climb out of our cars (mentally and physically) and try walking again. Remember your body? It's not a concept, it's not virtual reality. It's an animal. Animals 'R' Us, remember?

If technology isn't neutral, and using it isn't free, and beating it is difficult, maybe we need to redefine the problem. Instead of building a better mousetrap, can we ask whether we want to be trapping mice? Or ourselves? Can we turn off the machines in our own lives long enough to hear ourselves think and talk—and to hear our human neighbors think and talk, and to hear our non-human neighbors think and talk, and to hear the land itself thinking and talking—and reconsider what strange fruit we're making, and for whom it is made, and whether there might indeed be an alternative reality available, one that came before cynicism and surfeit?

Most of us most of the time see technology and technological progress as imperatives. Some of us are against technology, and some of us are all for it, but most of us are ambivalent as we slave away. Many people see the machine culture as an imperative, with a life and a will (virtual though they may be) of its own. But nuclear power plants and automobiles and polyvinylchloride are no more inevitable than an addiction to candy bars. A machine is not a living being; it has no will. It is as rational (blind), will-less, and disciplined (does what it's programmed to do) as Weber thought the bureaucrats to be. A machine is only as automatic as its computer program and its supply of fuel and its human addicts allow it to be. Do we want to be machines, blind and obedient, manufacturing and consuming plastic- and foil-wrapped artificially flavored candy that goes in one end and out the other without a trace of nutrition?

In the 1970s, the U.S. Senate Church Committee, after an investigation leading to shocking (to some) revelations of the U.S. government's illegal spying and institutionalized murder and mayhem, came up with ninety-six recommendations (and we forgive them for that, for they were after all lawmakers) ranging from Congressional and Justice Department oversight of intelligence agencies; to limiting those agencies' budgets; to forbidding intelligence agencies from domestic spying and covert activities unless there is suspicion of criminal activity; to not collecting unnecessary personal information about American citizens; to obtaining judicial warrants allowing domestic spying; to prohibiting the collection of information on the political beliefs, political associations, or private lives of Americans "except that which is clearly necessary for domestic security investigations"; to centralizing domestic security investigations under the FBI; and prohibiting the military, the National Security Agency, and the IRS from domestic spying.[9]

We have seen how centralization, secrecy, and rules are three inherent characteristics of bureaucracies. Unfortunately, the committee's recommendations were basically a call for more centralization and more rules.

A generation later, there is Congressional oversight over intelligence agencies, but wiretapping warrants are much easier to gain and domestic spying is much easier to do legally. When violations of civil liberties and privacy are disclosed, the best civil rights lawyers come up with virtually the same prescriptions: more oversight, more centralization, more rules. It seems the disease has progressed, but not the medicine.

So we might cast our jaundiced eyes beyond modern progress and attend to Lao-tzu, who warned that rules and structures were for people without responsibility and awareness, and that punishments arose only when people wouldn't govern themselves. He also pointed out that the more rules were promulgated, the more criminals there would be. Lao-tzu lived a long, long time ago, but he already saw the danger of device and recommended that people reduce their dependence on technology and stay close to home.[10]

A couple millennia later, Ralph Waldo Emerson lamented that "work and days were offered us, and we chose work."[11] We can choose days and save time. Gardens are for those who don't have time for supermarkets and all their consequences. Walking is for those who don't have time for driving. And there's more to gain than time, though we'll enjoy lots of that, too, when we return to the real world.

Just do it.

The late U.S. Senator J. William Fulbright wrote, "In a democracy, dissent is an act of faith."[12]

We write, In a machine, awareness is an act of freedom.

The controversies over privacy and security ought actually to be about technology and fascism. Privacy and security are worse than false hopes; they are distractions. We are going down the ecological and social tubes, so there is no security. And privacy never existed except in the fantasies of those who saw themselves as separate from their fellow beings.

The human nervous network is a complex of billions of neurons and synapses washed in biochemicals sending subtle multileveled messages. Human communities consist in part of millions of these nervous networks interacting. More-than-human communities consist in part of millions of species interacting in verbal and non-verbal ways. We can't fathom the world, much less control it, much less redesign it after our ambitions' impoverished goals. By trying to monitor everyone's behavior and thinking, and by trying to compel everyone to follow rules, we reveal our ignorance of the complexity and subtlety of human cultures and the natural world. Best to relax our obsessions with security and control, pay attention to what's going on in ourselves and others as best we can, and respond using our entire sensibilities and collective intelligence. A computer can intercept and record telephone and e-mail messages, but it cannot make sense of them. Computers are no substitute for intelligence. Rules are no substitute for human and nonhuman communities' evolved ways of interacting together. Closed circuit TV cameras are no substitute for social cohesion. Computers are no substitute for intelligence. Sense-making cannot be tacked together with a computer programmer's codes. Security cannot be had with weapons and spying. There is no such thing as intellectual property. You cannot protect your privacy. Obedience is never compelled. You cannot keep up with the machine's schedule of production. You cannot recycle industrial waste. You cannot create or replace a forest, a wetland, a human culture, a nonhuman culture, a language.

I often think about the religion of my youth and wonder what it would have been like to grow up knowing another sort of God, one who did not sit at the center of a universal Panopticon and sort us all: to heaven with this one and to hell with this other. One who did not wield power from a distance. One who did not live in the sky. One without an oversight that is always possible and (conveniently) never verifiable.

How would our culture be different if the sacred resided instead in our own bodies and in the bodies of those humans and nonhumans who are our neighbors and companions in the world? What sort of society would evolve if we understood that truth is multivaried and relative and just as manifest in dragonflies, phoebes, and drops of rain as in humans? How would our society look—and how would the world look—if our gods and our goddesses valued life over power? How would *we* look if *we* valued life over power? What would happen if we rejected the myth of the machine, and the machine itself?

Three stories.

The first appeared today in the *Los Angeles Times*. "There are no sounds from outside, because there are no windows—only a skylight high overhead, through which gray daylight seeps into the bare quadrangle facing the pod's eight cells, stacked four on four. All that can be heard are a few subdued voices, and the occasional thunderous sound of a flushing toilet reverberating off the blank concrete walls. This is not the crowded, clamorous kind of prison you see in the movies. The SHU, as it's known, is a starkly efficient place of electronically controlled doors and featureless concrete and steel. Occasionally, the monotony is punctured by bursts of noise and violence. Sometimes inmates scream at guards, other inmates, or themselves. Sometimes there is the clangorous racket of a recalcitrant prisoner being forcibly extracted from his cell. But most of the time, nothing happens. Almost nothing is permitted to happen. That's the idea of the SHU. . . . In the SHU, there are no jobs, no activities, hardly any educational programs and barely any human contact. You are locked in your 8-by-10-foot cell almost around the clock. You can't see the other prisoners in the cells adjoining yours, nor the guards watching from a central observation booth. Most of the time, all you can see through the fingertip-sized perforations in your cell's solid steel door is the wall of the eight-cell pod, the larger cage containing your cage. Guards deliver your meals. Once a day, the remote-controlled cell door grinds open, and you get 90 minutes to spend alone in a walled-in courtyard—a place more like the bottom of a mine shaft than an exercise yard. It's an environment about as restrictive and monotonous as human minds can design—and, perhaps, as human minds can tolerate."[13]

This is the quintessence—and future—of our society. This is the center of the machine.

Before you protest that you are not yourself inside of the SHU (and you would never be) look at your own surroundings. What walls surround you? Do you live in the real physical world outside of the machine?

A major difference between those in the SHU and the rest of us is that at least they have the excuse of locked doors between them and the real world. What is our excuse?

Another story.

I have before me several pictures of a "happy event" that took place last year at Plymouth Rock, where the civilized first brought their panoptic god to the people and landscape of what is now Massachusetts. The photographs show hundreds of American Indians meeting with Christians, so that the Indians could get down on their knees and apologize for not accepting Jesus when the white people first offered. Plymouth is not the only place Indians will be allowed to kneel and apologize. The Christian ministers behind this are now taking their show on the road, giving Indians all over North America the opportunity to apologize because their ancestors did not quickly enough surrender themselves to the machine.

This guilt and redemption too is the quintessence—and future—of our society. It is the more or less final incorporation of wild and free individuals into the machine. Not only must we put up with the imposition of the Panopticon and its theft of all our inheritance, but we must apologize for not giving up our lives sooner. We must be thankful for all it has done to us, and be redeemed by the machine itself, for it is God. It is the movement from "Thou shalt not" to "Thou shalt" and finally to "Thou art."

A third story, from the Oglala Red Cloud: "Friends, it has been our misfortune to welcome the white man. We have been deceived. He brought with him some shining things that pleased our eyes; he brought weapons more effective than our own; above all, he brought the spirit water that makes one forget for a time old age, weakness, and sorrow. But I wish to say to you that if you would possess these things for yourselves, you must begin anew and put aside the wisdom of your fathers. You must lay up food, and forget the hungry. When your house is built, your storeroom filled, then look around for a neighbor whom you can take at a disadvantage, and seize all that he has! Give away only what you do not want; or rather, do not part with any of your possessions unless in exchange for another's.

"My countrymen, shall the glittering trinkets of this rich man, his deceitful drink that overcomes our mind, shall these things tempt us to give up our homes, our hunting grounds, and the honorable teaching of our old men? Shall we permit ourselves to be driven to and fro—to be herded like the cattle of the white man?"[14]

This, too, is the quintessence of the future. This is the quintessence of resistance to the machine. This is a statement of what it means to be alive.

And one more story, a confession from someone in the inner ring of the Panopticon. A high-ranking security chief from South Africa's apartheid regime later told an interviewer what had been his greatest fear about the rebel group African National Congress (ANC). He had not so much feared the ANC's acts of sabotage or violence—even when these were costly to the rulers—as he had feared that the ANC would convince too many of the oppressed majority of Africans to disregard "law and order." Even the most powerful and highly trained "security forces" in the world would not, he said, have been able to stem that threat.[15] As soon as we come to see that the edicts of those in power are no more than the edicts of those in power, that they carry no inherent moral or ethical weight, we become the free human beings we were born to be, capable of saying *yes* and capable of saying *no*.

Remember that.

Éttiene de La Boétie reminded us long ago that when the powerful are insatiable, submission is fatal, that the more we submit ourselves to the "law and order" of those in power, the more they will demand. He wrote that "the more tyrants pillage, the more they crave, the more they ruin and destroy; the more one yields to them, and obeys them, by that much do they become mightier and more formidable, the readier to annihilate and destroy. But if not one thing is yielded to them, if, without any violence they are simply not obeyed, they become naked and undone and as nothing, just as, when the root receives no nourishment, the branch withers and dies."[16]

In his powerful *Modernity and the Holocaust,* the sociologist Zygmunt Bauman put this in a slightly different way, stating that the terror by which the Panopticon is maintained only "remains effective as long as the balloon of rationality has not been pricked. The most sinister, cruel, bloody-minded ruler must remain a staunch preacher and defender of rationality—or perish. Addressing his subjects, he must 'speak to reason.' He must protect reason, eulogize on the virtues of the calculus of costs and effects, defend logic against passions and values which, unreasonably, do not count costs and refuse to obey logic."[17]

In their film *The Matrix,* the Wachowski brothers, producers and directors, powerfully described some aspects of our culture. Early in the film we see people walking through their days pretty much as we do ours. But that "ordinary" life was an illusion, a computer program, and the machines running this program were actually using energy extracted from human bodies. It was an extraordinarily clear and powerful metaphor for the workings of the machine.

But the brothers also got part of it wrong (besides the unfortunately simple cure of taking a pill). In the film, when you leave the matrix you find yourself surrounded by a drab, ruined world, by ugliness, by people still relying on machines, only less sophisticated ones. The problem here is that the filmmakers have confused the loss of some physical comfort and the illusion of emotional security—both of which the machine does provide for some—with the loss of beauty and of wildness, neither of which the machine provides anyway.

That is a huge mistake. Let me speak for myself. In leaving the matrix, the Panopticon, the megamachine, the system—all of these modes of thought that enslave us—I have found myself surrounded not by drabness but by ecstatic beauty. To provide one example among scores each day, I worked late last night, hour after hour in front of the computer, writing. But when I left the machine I went outside and saw more beauty than I could create in a thousand lifetimes of writing, more beauty than the machine will be able to manufacture before it all collapses. Thin clouds ran in front of a moon sharp and bright enough to make the clouds translucent. Near the ground, the air was still. The tops of trees did not move. A bat fluttered in the partial light. In the distance I heard the barking of sea lions.

We can see ecstatic life anywhere. We pretend that the natural world does not exist in cities, but it does. I always stroke the trees I see there, let them know I care, and that I am sorry for their imprisonment in concrete and brick and asphalt. Even people who live in cities can still find nonhumans with whom they can interact, and to whom they can listen. There are ants and spiders. There are birds. There are plants. I love to watch and encourage and thank the plants I see pushing up sidewalks. That is work we should all be doing, and they are leading the way, teaching us how. These plants reaching through the concrete from the soil to the sky, these ants and birds and spiders going about their lives, all remind us that at all times and in all places—even in cities—ecstatic life continues beneath the machine, waiting for the chance to return, to recover, and to reenter into relationship with those of us who are ready to live.

That said, to leave the matrix entirely, to enter places not yet too ground up by the machine, is to remember what it is to be alive. The world smells different in a living forest. Time changes meaning. Sounds change texture. To listen to the hollow booming of a pileated woodpecker is to be reminded that we do not inhabit a world ruled by computers and two-cycle engines, that these minor artifacts are peripheral to our lives, that our real home is the wild, that we ourselves are wild.

I do not need a heaven, either technological or religious.

This life is good enough.

The Wachowski brothers were making a Hollywood movie, after all, but the truth is known to every wild human, known to every crane fly and every willow, every amanita mushroom and every bear.

This life is good enough.

I'll tell you a secret, a secret so dirty that the machine and its servants must make sure to wash our brains daily so we do not pollute ourselves or the machine with its filth, its insanity, its blasphemy.

Are you ready?

The distant sky God, the guard at the center of the Panopticon, governments, and corporations have much in common. They are all about centralizing power, all about control. They are all about converting the living to the dead because only the dead can be controlled.

But there is something else the power-gods hold in common. This is the secret: they do not exist. The machine itself does not exist. It, like all of these other social constructs, has consequences, to be sure, but it does not really exist. We pretend it does, but our pretensions do not make it real.

Privacy does not exist. I am not alone. I can never be alone. I do not exist independent of everyone else. I affect those others, and they affect me. I communicate with them, and they communicate with me.

Property does not exist. I do not own the chair I am sitting on, nor the cat that sits next to me. I do not own the land where I live. I am sitting on a chair we all agree belongs to me, next to a cat we all agree belongs to me, on land we all agree belongs to me. Weyerhaeuser does not own land. Freeport McMoRan does not own land. No corporation owns land. We all agree that they own land. And so we let them destroy our landbases. That's not very smart. It's also unnecessary. It takes only one person in a relationship to change a relationship, though change may mean the relationship ends.

I'll tell you another secret. We may be spending our days and nights in the center of the machine, working for this thing that does not exist to get things that don't make us happy. We may be living in the outer ring of this Panopticon that does not really exist. But unlike those in the SHU, we can walk away. We can deliberately refuse the bargain offered by the machine, refuse to give up the richness of our own lives and the lives of those we love for shining things that please our eyes, for protection from their weapons that are more effective than our own, and above all for the spirits of the machine that make us forget for a time old age, weakness, and sorrow. We can go on total strike against the terms offered by the machine.

Here is the secret: we can say *no* to the machine.

And we can say *yes* to our own lives, and to the lives of those we love.

My students at the prison told me that plastered all over the walls of the L.A. County Jail, and across the walls of jails everywhere, is the phrase "Make a Deal." Many of my students told me that one of the easiest, most direct ways to shut down the whole court system would be for prisoners to stop making it easier on their captors, to not make deals.

We are not deceived that it is always easy to say *no* to the machine. Although it does not exist, there are those who believe in it enough that they may imprison you, they may torture you, they may kill you. This is true. There are few of us who are naïve enough to believe otherwise.

But here's one more secret. Until those in power figure out a way to completely replace humans with machines, there will always be more of us than there are of them. All it will take for this whole rotten system to collapse is for enough of us to learn to say *no*. And to say *no* again. And again. And again. And again.

And for enough of us to learn to say *yes*.

Do not be a tourist in your own life, your own body, your own landscape, your own world. Remember, this is where you *live*. This is your life. Live it wisely. Live it in service to those you love, not in service to the machine.

You walk into a store. Or maybe it's a police station. Or a school. Sometimes you can't tell the difference. You know that there are cameras everywhere. You can see them. You know that there are no chips in your clothes. You have removed them, each and every one.[18] You know that someday the world will once again be free of machines, and free of the machine culture that made them. But for now it isn't.

And you aim to do something about that.

You're not stupid. You know that there are many who still believe in the myth of the machine. You know that there are many who are more interested in power than life—yours or theirs. You know that they will kill to defend the machines that own them. You know that they will themselves die rather than break their identification with the machine. In fact the machine is already killing them. And you will not be caught by them. You are not stupid.

You do not take orders from anyone. Sometimes you work by yourself, and sometimes you work with those you trust, your brothers and sisters in this struggle against the machine, for life. You are dismantling the machines, and the machine, in the ways you know best. You have taken them out of your heart, and glittering trinkets no longer hold you captive. You see the store, or maybe it's a police station, or a school, or a political organization, for the prison that it really is.

And when you are done with your work inside of this prison, dismantling whatever needs dismantling that you are best able to dismantle in the way that you are best able to dismantle it, you go back outside, into the sunshine or the dark of night, into the embrace of the real world.

You feel happy.

Acknowledgments

Derrick

A couple of years ago, I received in the mail a 'zine from someone named Nick Evans. The 'zine potently explored the capacity of the government (any government, but he focused on the United States) to surveil us and even to control our thoughts. It was horrifying, but there was just one problem: the things Evans described were frankly unbelievable. Nobody can read my mind, I thought. And the only way someone can inject thoughts into my brain is through good old-fashioned indoctrination, not microwaves. His 'zine contained the stuff of science fiction.

I traced Nick's sources, fully expecting them to be crazy and lack credibility. I was wrong. He cited original sources, including peer-reviewed journals and Department of Defense Web sites (okay, so the DoD is crazy and has zero credibility, but we'll leave that aside). What he described is real. I knew immediately I wanted to explore this subject.

So my first thanks are to Nick for having the courage and intelligence to act on this very real danger. This book would not have been written were it not for him.

I would also like to thank the land where I live for teaching me that DARPA is wrong, that knowledge isn't power, that knowledge is knowledge, and knowledge is intimacy. This book could not have been written without the help of this land.

I am grateful to George Draffan for his extraordinary work. Once again, this book could not have been written without him.

Special thanks to Ian Baldwin, Michael Baldwin, Philip Baldwin, Joan Baez, Karen Breslin, MaryAlecia Briggs, Rebecca Bruce, Gabrielle Benton, Leha Carpenter, Nita Crabb, Bill Gresham, Mary Gresham, Louis Hazard, Mary Jensen, Lichen

June, John Fish Kurmann, Casey Maddox, Remedy, Tiiu RFT Ruben, Julianne SkaiArbor, Becky Tarbotton, Karen Tweedy-Holmes, Becca Wissbaum.

Finally, I would like to thank Anthony Arnove for agenting, Helen Whybrow for editing, Lori Lewis for copyediting, Margo Baldwin for publishing this book, and Collette Leonard and Marcy Brant for overseeing the whole process.

George

I would like to thank Jeremy Bentham, for his prescient revelations about the intentions and capabilities of the men who run the machine. And thanks to the Panopticon's security guards, for their arrogant disregard for common sense and for their violations of decency, which periodically rouse us from our slumber. Thanks to the spirit of William Burroughs, for pointing out that a paranoid is someone who has only just figured out what's going on, and for revealing the gulf between conspiracy theories and facts. And to John Schinnerer at Eco-Living.net, for being the wizard who kept my computer going long enough to write another book attacking technology.

Notes

The All-Seeing Eye

1. The "debates" rage around whether or not it's a problem, whether the loot is being fairly distributed, whether the benefits have been worth the destruction, and what if anything can or should be done to mitigate the damage.

2. Carey Goldberg, "Some Fear Loss of Privacy as Science Pries into Brain," *Boston Globe,* May 1, 2003, sec. A.

3. Faye Flam, "Your Brain May Soon Be Used Against You," *Philadelphia Enquirer,* October 29, 2002, http://www.philly.com/mld/inquirer/4391614 .htm.

4. Patrick Suppes, Bing Han, and Zhong-Lin Lu, "Brain-wave Recognition of Sentences," *Proceedings of the National Academy of Sciences* 95, no. 26 (December 22, 1998), 15861–66, http://www.pnas.org/cgi/content/full/95/ 26/15861.

5. Katherine Albrecht, "Auto-ID, Tracking Everything, Everywhere," from "Supermarket Cards: Tip of the Retail Surveillance Iceberg," *Denver University Law Review* 79, no. 4 (Summer 2002).

6. Auto-ID Center, "The Future of Business," http://www.autoidcenter.org/ aboutthecenter.asp (accessed August 14, 2003).

7. Auto-ID Center, "Sponsor Companies," http://www.autoidcenter.org/ sponsors_companies.asp (accessed November 5, 2003).

8. Cheryl Rosen and Mathew G. Nelson, "The Fast Track," *InformationWeek,* June 18, 2001, http://www.informationweek.com/ shared/printableArticle?doc_id=IWK20010618S0001.

9. Winston Chai and Richard Shim, "Benetton Considers Chip Plans," *CNET News*, April 7, 2003, http://news.zdnet.co.uk/hardware/chips/ 0,39020354,2133031,00.htm.

10. Albrecht, "Auto-ID, Tracking Everything, Everywhere."

11. John Gartner, "Point 'n' Shoot Sound Makes Waves," *Wired News,* February 21, 2002, http://www.wired.com/news/technology/0,1282,50483 ,00.html (accessed May 20, 2003).

12. Cheryl Welsh, "Best Mind Control Documentary Excerpts," http://www.dcn.davis.ca.us/~welsh/tvlist.htm (accessed May 20, 2003).

13. Rodger D. Hodge, Weekly Review, *Harper's Magazine,* June 3, 2003.

14. R. D. Laing, *The Politics of Experience* (New York: Ballantine Books, 1977), 58.

15. Jeremy Bentham, *Panopticon; Or The Inspection-House: Containing The Idea Of A New Principle Of Construction Applicable To Any Sort Of Establishment, In Which Persons Of Any Description Are To Be Kept Under Inspection; And In Particular To Penitentiary-Houses, Prisons, Houses Of Industry, Work-Houses, Poor-Houses, Lazarettos, Manufactories, Hospitals, Mad-Houses, And Schools: With A Plan Of Management,* http://cartome.org/panopticon2.htm (accessed June 10, 2003).

16. Bentham, *Panopticon.*

17. Bentham, *Panopticon.* Extravagant italicization in original.

18. Bentham, *Panopticon.* Bentham's use of the term "Purpose X" is revealing. The purpose of surveillance and exact mandate of the bureaucracy doing it are never made explicit. The rationales are invariably abstract virtues such as taxpayer value, customer service, or the ever-popular "security," which covers everything from happy retirement to preemptive wars. The U.S. government slipped of course when it began to publicly equate national security and economic security; for a discussion of the Carter Doctrine and its informal predecessors, see Michael T. Klare, *Resource Wars: The New Landscape of Global Conflict* (New York: Metropolitan Books, 2001), 33ff.

19. Michel Foucault, *Discipline & Punish: The Birth of the Prison,* trans. Alan Sheridan (New York: Vintage Books, 1979), 201–202.

20. Foucault, *Discipline & Punish,* 202.

21. George Washington, First Inaugural Address (April 30, 1789), published at *The Avalon Project at Yale Law School,* http://www.yale.edu/lawweb/avalon/presiden/inaug/wash1.htm (accessed June 14, 2003).

22. Nathan Miller, *Stealing from America: A History of Corruption from Jamestown to Reagan* (New York: Paragon House, 1992), 56–62.

Science

1. Jane Wakefield, "US Looks to Create Robo-soldier," *BBC News,* April 10, 2002, http://news.bbc.co.uk/1/hi/sci/tech/1908729.stm (accessed June 21, 2003).

2. Guy Gugliotta, "Rats Turned into Remote-Controlled Robots: Techniques Potential Uses Include Aid to Victims of Disaster or Neural Injuries," *Washington Post,* May 2, 2002, sec. A, http://www.washingtonpost.com/ac2/wp-dyn?pagename=article&node=&contentId=A18261-2002May1&notFound=true.

3. Colin A. Ross, *Bluebird: Deliberate Creation of Multiple Personality by Psychiatrists* (Richardson, Tex.: Manitou Communications, 2000), 97.

4. David Hearst, "Sci-fi War Put under the Microscope," *The Guardian,* May 20, 2003, http://www.guardian.co.uk/uk_news/story/0,3604,959469,00.html (accessed June 21, 2003).

5. Hearst, "Sci-fi War."

6. Associated Press, "Pentagon Hopes to Identify People by the Way They Walk," *News-Star* (Shawnee, Okla.), May 20, 2003, http://www.news-star .com/stories/052003/New_8.shtml.

7. Michael J. Sniffen, "Proposed System Would Use Lots of Data," *The Guardian,* May 19, 2003.

8. Francis Bacon, *The Works,* ed. James Spedding (New York: Garrett Press, 1968), 4:296, quoted in Carolyn Merchant, *The Death of Nature: Women, Ecology and the Scientific Revolution* (San Francisco: HarperSanFrancisco, 1983), 164, 168, and 172.

9. It's biochemical warfare: sticking with your habitual responses (for most of us, pleasure, comfort, safety, orientation around a tight sense of self) produces opiumlike substances in the brain and eventually sets the patterns of the brain's neural connections themselves. Breaking out of habitual patterns results in uncomfortable biochemical withdrawal. See the November 1999 issue of *Science* magazine and Edward M. Eveld, "It's Time to Break Those Bad Habits: Here's How," *Kansas City Star,* December 29, 2000, http://www.kcstar.com/item/pages/story,local/37750489.c29.html.

10. Francis Fukuyama, *The End of History and the Last Man* (New York: Free Press, 1992).

Control

1. Michael J. Sniffen, "Pentagon Developing System that Can Track Every Vehicle in a City," *CNEWS,* July 1, 2003, http://www.canoe.com/CNEWS/ TechNews/2003/07/01/124436-ap.html.

2. Deborah Bulkeley, "Webcam Surveillance in Biloxi Classrooms: Cameras Watch All Students in School District," *San Francisco Chronicle,* August 13, 2003.

3. Julia Scheeres, "The R's: Reading, Writing, RFID," *Wired News,* October 24, 2003, http://www.wired.com/news/technology/0,1282,60898,00.html ?tw=wn_techhead_5.

4. Andrew Cawthorne, "U.S. Raid Herds Iraqi Old and Young in Barbed Wire," *Excite News,* September 1, 2003, http://news.excite.com/world/

article/id/51496|world|09-01-2003::01:48|reuters.html (accessed September 2, 2003).

5. *CNN News,* "Police, School District Defend Drug Raid," *CNN.com,* November 7, 2003, http://www.cnn.com/2003/US/South/11/07/school.raid.

6. Edward Abbey, letter to the editor, *Bloomsbury Review,* April 1985.

7. Sunshine Project, "Ethnically-specific Biological Weapons: An Analysis of Human Genome Data Reveals That Ethnically-specific Genetic Markers Do Exist," briefing paper from *Emerging Technologies: Genetic Engineering and Biological Weapons,* Sunshine Project Backgrounder No. 12, October 2003, http://www.sunshine-project.org/publications/others/snpbw.html (accessed October 25, 2003).

8. Sunshine Project, "Ethnically-specific Biological Weapons."

9. Thomas Donnelly, "Rebuilding America's Defenses: Strategy, Forces and Resources," the Project for the New American Century, September 2000, http://www.newamericancentury.org/RebuildingAmericasDefenses.pdf.

10. As described, for example, by R. D. Laing, *The Politics of Experience.*

11. How different is targeting specific genotypes from the Tuskegee study? After all, syphyllis has been used as a biological weapon.

12. Donnelly, "Rebuilding America's Defenses."

13. U.S. Defense Advanced Research Projects Agency (DARPA), "Harvesting Biology for Defense Technology Conference, June 23-25, 2003," card at http://web-ext2.darpa.mil/body/ppt/DARPAPostCARD.PPT (accessed September 1, 2003).

14. Cheryl Seal, "Frankensteins in the Pentagon: DARPA's Creepy Bioengineering Program," *Information Clearinghouse: News You Won't Find on CNN,* http://www.informationclearinghouse.info/article4572.htm (accessed September 1, 2003).

15. Seal, "Frankensteins in the Pentagon."

16. Erik Baard, "The Guilt-Free Soldier: New Science Raises the Specter of a World Without Regret," *The Village Voice,* January 22–28, 2003, http://www.villagevoice.com/issues/0304/baard.php (accessed September 2, 2003).

17. Kinder Investigations & K-9 Drug Detection Services, "Amphetamines," http://www.k9investigations.com/amphetamines.htm (accessed September 2, 2003).

18. U.S. Air Force Scientific Advisory Board, *New World Vistas: Air and Space Power for the 21st Century* (Washington, D.C.: USAF Scientific Advisory Board, 1995–96), 15:89, http://stinet.dtic.mil/Cgibin/fulcrum_main.pl

?database=ft_u2&searchid=10677104835631&keyfieldvalue=ADA309597
&filename=%2Ffulcrum%2Fdata%2FTR_fulltext%2Fdoc%2FADA309597
.pdf.

Identity

1. Bob Sullivan, "Why We're All at Risk of ID Theft," *MSNBC,* January 21, 2003, http://www.msnbc.com/news/758896.asp (accessed September 5, 2003).

2. Jennifer Kerr, "Group Gets Private Data on U.S. Officials," Associated Press, August 28, 2003, http://story.news.yahoo.com/news?tmpl=story&u=/ap/20030828/ap_on_hi_te/privacy/concerns_5.

3. The following statistics are all from U.S. Federal Trade Commission, *Identity Theft Report* (McLean, Va.: Synovate, September 2003), 4, 6–7, 12, 28, 30-31, 50, 54, 59, http://www.ftc.gov/os/2003/09/synovatereport.pdf.

4. Sullivan, "Why We're All at Risk."

5. Kerr, "Group Gets Private Data."

6. BBC News, "Health: Breast Milk Studied for Toxins," *BBC Online Network,* July 12, 1999, http://news.bbc.co.uk/1/hi/health/391514.stm (accessed November 5, 2003).

The Machine

1. As one woman wrote me: "I want to comment on how sad the statistics cited in both your books on the rate of women raped in our country are to me. Not because they are so high, but because they are so completely inaccurate (I know you realize this). My personal experience and that of ALL the women I love has been much different in that we have ALL been raped, even by the miserably incomplete statutory definition of rape in Washington state. Of all the cases of rape I know of personally (seven), only one was reported to the police. Through mishandling by the police and school district and a complete disregard for the victim, nothing happened to the rapist. My point is that the statistics of 25 percent raped and 19 percent attacked completely misrepresent the reality of being a woman in this world. The real statistic is 100 percent: if you are a female, you will be raped. And while we're at it, let's write a statutory definition of rape that is accurate, something like 'acting on the objectification of the feminine.' This definition would cover the earth, children, men (in that we are all male and female beings and the part of men that is raped is their feminine, the 'lesser' part), anything and everything. Sadly, the rape of our planet and everything connected to it will continue as long as objectification for any sake—production, the White male, the masculine—is still accepted and encouraged."

2. George Orwell, *1984* (New York: Signet Classics, 1962), 220.

3. Orwell, *1984,* 220.

4. Andy Sullivan, "Cutting-Edge Science Creates Stain-Free Pants," *USA Today,* July 23, 2003, http://www.usatoday.com/tech/news/techinnovations/2003-07-23-robopants_x.htm.

5. Charles Platt, "The Museum of Nanotechnology," *Wired,* http://www.wired.com/wired/scenarios/museum.html (accessed August 17, 2003).

6. William Prendergast, nanotechnology patent lawyer and partner in the Chicago law firm of Brinks, Hofer, Gilson & Lione, August 28, 2003, http://www.etcgroup.org/main.asp (accessed October 22, 2003).

7. ETC Group, *The Big Down: Atomtech: Technologies Converging at the Nano-scale* (Winnipeg, Manitoba: ETC Group, 2003), 12, http://www.etcgroup.org/documents/TheBigDown.pdf.

8. The technology critic is the ETC Group's Jim Thomas, here quoted by the technology apologist Noah Shachtman in "Rage Against the (Green) Machine," *Wired News,* June 19, 2003, http://www.wired.com/news/technology/0,1282,59287,00.html.

9. For example, Michael Crichton's *Prey.*

10. Betterhumans, "Immortality," November 20, 2002, http://www.betterhumans.com/Resources/Goals/goal.aspx?articleID=2002-05-08-5 (accessed September 8, 2003).

11. Betterhumans, "Immortality."

12. Betterhumans, "Immortality."

13. Ray Kurzweil, "We Are Becoming Cyborgs," *KurzweilAI.net,* http://www.kurzweilai.net/meme/frame.html?main=/articles/art0449.html (accessed September 8, 2003). In the 1950s, William S. Burroughs wrote stuff like this as satire. Now it's multibillion-dollar cutting-edge science. I guess this would invert Marx's line about history repeating itself, first as tragedy, and then as farce. This started off as farce and has moved quickly toward tragedy.

14. Bruce J. Klein, "Building a Bridge to the Brain: Researchers Are Close to Breakthroughs in Neural Interfaces, Meaning We Could Soon Mesh our Minds with Machines," Betterhumans, March 3, 2003, http://www.betterhumans.com/Features/Reports/report.aspx?articleID=2003-03-02-3. Interestingly enough, when I visited this Web site, the sponsoring advertisement (for Yahoo) said, "Be careful what you wish for."

15. Alexander Bolonkin, "Science, Soul, Heaven and Supreme Mind: Russian Scientist Alexander Bolonkin Develops Artificial Intelligence in the USA,"

Pravda, January 8, 2003, http://english.pravda.ru/main/2003/01/08/41749 .html. With all of Bolonkin's quotes, the awkwardness is in the original. His first language is Russian.

16. Alexander Bolonkin, "Twenty-First Century—The Beginning of Human Immortality," http://bolonkin.narod.ru/Bolonkin-p3.htm (accessed September 9, 2003).

17. Bolonkin, "Twenty-First Century."

18. Bolonkin, "Twenty-First Century."

19. Bolonkin, "Science, Soul, Heaven."

20. Orwell, *1984,* 220.

21. Bolonkin, "Science, Soul, Heaven."

22. Bolonkin, "Twenty-First Century."

23. Bolonkin, "Science, Soul, Heaven."

24. *Rachel's Environment and Health News,* "The Revolution. Part 1," June 26, 2003, http://www.rachel.org/bulletin/bulletin.cfm?Issue_ID=2362.

25. Lewis Mumford, "Drama of the Machines," *Scribner's Magazine,* August 1930, in *Interpretations and Forecasts: 1922–1972* (New York: Harcourt Brace Jovanovich, 1973).

26. K. Eric Drexler, "Engines of Destruction," in *Engines of Creation: The Coming Era of Nanotechnology,* Foresight Institute, http://www.foresight .org/EOC/EOC_Chapter_11.html (accessed September 10, 2003).

27. K. Eric Drexler, "Foresight Background 3, Rev. 1: Dialog, Exploratory Engineering, Bioarchive," Foresight Institute, http://www.foresight.org/ Updates/Background3.html#DangerDialog (accessed September 10, 2003).

28. *Rachel's,* "Part 1."

29. *Rachel's Environment and Health News,* "The Revolution. Part 3," July 24, 2003, http://www.rachel.org/bulletin/bulletin.cfm?Issue_ID=2371.

30. *Rachel's Environment and Health News,* "The Revolution. Part 2," July 10, 2003, http://www.rachel.org/bulletin/bulletin.cfm?Issue_ID=2363.

31. *Rachel's,* "Part 2."

Fear

1. There have been studies that estimate that the average person thinks a thousand thoughts per hour. That's twelve thousand thoughts per day, and really busy thinkers might have fifty thousand thoughts in a day. How many of these thoughts are mere repetition? Obsession has been defined as "persistent ideas, thoughts, impulses, or images that are experienced as

intrusive and inappropriate and that cause marked anxiety or distress."
See Actualizations, "Obsessive Compulsive Disorder," http://www
.actualizations.com/ocd.htm#what%20is (accessed October 27, 2003). So if
you don't *feel* the anxiety or impropriety, then your repetitive thoughts
about, say, buying this or that product aren't *obsessive.*

2. Eduardo Galeano, *Upside Down: A Primer for the Looking-Glass World,*
   trans. Mark Fried (New York: Picador USA, 2001), 80.

3. Vine Deloria, "Where The Buffalo Go: How Science Ignores the Living
   World: An Interview with Derrick Jensen," *The Sun,* July 2000.

4. Marc Smith, "Church History: Kill Them All, Let God Sort Them Out!"
   Straitway, http://straitway.org/2001/03012001.htm (accessed September 15,
   2003).

5. Regine Pernoud, *Those Terrible Middle Ages!* Translated from the 1977
   French edition by Anne Englund Nash (San Francisco: Ignatius, 2000), 19.

Rationalization

1. This is happening in California right now.

2. James Beniger, *The Control Revolution: Technological and Economic Origins
   of the Information Society* (Cambridge, Mass.: Harvard University Press,
   1986), 15, cited in Oscar H. Gandy Jr., *The Panoptic Sort: A Political
   Economy of Personal Information* (Boulder, Colo.: Westview Press, 1993), 42.

3. The following quotes are all from Karl Marx and Friedrich Engels,
   Manifesto of the Communist Party, 1848, http://www.marxists.org/archive/
   marx/works/1848/communist-manifesto/.

4. The authors are indebted to Frank Elwell's discussion of Weber in "The
   Sociology of Max Weber," http://www.faculty.rsu.edu/~felwell/Theorists/
   Weber/Whome.htm (accessed September 16, 2003).

5. Max Weber, *The Protestant Ethic and the Spirit of Capitalism,* trans. Talcott
   Parson (New York: Charles Scribner's Sons, 1904, 1930), 60.

6. Weber, *The Protestant Ethic,* 157–158.

7. Max Weber, *From Max Weber,* trans. and ed. H. H. Gerth and C. Wright
   Mills (New York: Galaxy, 1946, 1958), 172.

8. Max Weber, *Economy and Society,* ed. Guenther Roth and Claus Wittich,
   trans. Ephraim Fischoff and others (New York: Bedminster Press, 1921,
   1968), 1156.

9. Robert Michels, *Political Parties: A Sociological Study of the Oligarchical
   Tendencies of Modern Democracy,* trans. Eden Paul and Cedar Paul (New
   York: The Free Press, 1915), 401.

10. Weber, *From Max Weber,* 229.

11. Weber, *Economy and Society,* 223.

12. Weber, *Economy and Society,* liii.

13. Max Weber, in J.P. Mayer's *Max Weber and German Politics,* 2nd ed. (London: Faber and Faber, 1956), 126–127. Cited online at *Verstehen: Max Weber's Home Page: A Site for Undergraduates,* Frank W. Elwell (Rogers State University), http://www.faculty.rsu.edu/~felwell/Theorists/Weber/Whome.htm.

14. Weber, *Economy and Society,* 224.

15. Weber, *From Max Weber,* 214.

16. Weber, *From Max Weber,* 214.

17. Weber, *From Max Weber,* 128.

18. Weber, *The Protestant Ethic,* 181.

19. Raul Hilberg, *Destruction of the European Jews,* revised and definitive edition (New York: Holmes & Meier, 1985), 3:1024.

20. Christopher R. Browning, *Fateful Months: Essays on the Emergence of the Final Solution* (New York: Holmes & Meier, 1985), 66–67, quoted in Zygmunt Bauman, *Modernity and the Holocaust* (Ithaca, N.Y.: Cornell University Press, 1992), 195.

21. C. Wright Mills, *The Causes of World War Three* (London: Secker & Warburg, 1958), 83–84.

22. U.S. Central Intelligence Agency, *KUBARK Counterintelligence Interrogation Manual,* July 1963.

23. Lewis Mumford, "Marx: Dialectic of Revolution," in *Interpretations and Forecasts: 1922–1972* (New York: Harcourt Brace Jovanovich, 1973), 202.

24. Lewis Mumford, *The Pentagon of Power* (New York: Harcourt Brace Jovanovich, 1970), 127.

25. Mumford, *The Pentagon of Power,* 127.

26. Mumford, *The Pentagon of Power,* 179.

27. Mumford, *The Pentagon of Power,* 178.

28. Mumford, *The Pentagon of Power,* 180.

29. Mumford, *The Pentagon of Power,* 208.

30. Mumford, *The Pentagon of Power,* 213.

31. Mumford, *The Pentagon of Power,* 434–435.

32. Mumford, *The Pentagon of Power,* 260.

33. Lewis Mumford, "The Origins of War," in *Interpretations and Forecasts: 1922–1972* (New York: Harcourt Brace Jovanovich, 1973), 333.

34. Mumford, *The Pentagon of Power,* 434–435.

The Panoptic Sort

1. Gandy, *The Panoptic Sort,* 15.

2. From a seventeenth-century order for dealing with the plague, quoted in Foucault, *Discipline & Punish,* 195.

3. Leo Lucassen, "A Many-Headed Monster," in *Documenting Individual Identity: The Development of State Practices in the Modern World,* ed. Jane Caplan and John Torpey (Princeton, N.J.: Princeton University Press, 2001), 237ff.

4. For example, in the 1940s Stalin moved nearly a million Chechen people based on the industrial state's need for labor; a quarter of them perished by the time they were allowed to return in 1957. Marc Garcelon, "Colonizing the Subject," in *Documenting Individual Identity,* ed. Caplan and Torpey, 98.

5. Caplan and Torpey, eds., *Documenting Individual Identity,* 11–12.

6. Act of November 10, 1919, Public Law 79, 66th Cong., 1st sess., vol. 104 (November 10, 1919).

7. Jon Agar, "Modern Horrors," in *Documenting Individual Identity,* ed. Caplan and Torpey, 119.

8. Modern ingenuity keeps up with the age-old art of changing one's identity. "Learn how to create a completely new identity. Get a fresh start. Establish a new credit file." (www.ezdiscountstore.com/newyou.html). "Change Your Identity without fake ID. Create an entirely new identity complete with a genuine new birth certificate, driver's license, social security card, new social security number, passport and even major credit cards!" (www.ariza-research.com/new-id).

9. Jane Caplan, "This or That Particular Person," *Documenting Individual Identity,* ed. Caplan and Torpey, 65.

10. Martine Kaluszynski, "Republican Identity," in *Documenting Individual Identity,* ed. Caplan and Torpey, 134–135.

11. American-Israeli Cooperative Enterprise, "Jewish Virtual Library," http://www.us-israel.org/jsource/Holocaust/Hollerith.html (accessed August 16, 2003), citing United States Holocaust Memorial Museum, http://www.ushmm.org/.

12. Wow! The U.S. biometric systems at the Mexican border are claimed to be accurate enough to distinguish between identical twins—just like their mothers, fathers, and friends do.

13. Jennifer Lee, "Passports and Visas to Add High-Tech Identity Features," *New York Times,* August 24, 2003, http://www.nytimes.com/2003/08/24/national/24IDEN.html.

14. Lee, "Passports and Visas."

15. Lee, "Passports and Visas."

16. Lee, "Passports and Visas."

17. Jeremy Bentham, "Principles of Penal Law," in *The Works of Jeremy Bentham,* ed. John Bowring (Edinburgh: William Tait, 1843), 1:557.

18. U.S. CIA, *KUBARK.*

19. "We" in this case doesn't mean George and Derrick, who have never tracked a gypsy in our lives. We mean the big We, the State. Or corporations. Insofar as there is a difference.

20. Connecticut Department of Social Services, "DSS's Biometric ID Project: Project Overview," http://www.dss.state.ct.us/digital/project.htm (accessed August 16, 2003).

21. Connecticut Department of Social Services, "DSS's Biometric ID Project: Connecticut Survey of Clients Who Have Just Been Digitally Imaged for AFDC 3/29/96 to 4/3/96," http://www.dss.state.ct.us/digital/faq/disurvey.htm (accessed August 16, 2003).

22. Connecticut Department of Social Services, "DSS's Biometric ID Project: Understanding Public Perception," http://www.dss.state.ct.us/digital/faq/disuppt.htm (accessed August 16, 2003).

23. Connecticut Department of Social Services, "DSS's Biometric ID Project: Understanding Public Perception."

24. Press Association, "DNA Database Stores 2m Profiles," *The Guardian* (UK), June 25, 2003, http://www.guardian.co.uk/crime/article/0,2763,984753,00.html (accessed August 14, 2003).

25. U.S. Congress, Counterterrorism Information Sharing with Other Federal Agencies and with State and Local Governments and the Private Sector: Hearing before Joint House/Senate Intelligence Committee, October 1, 2002 (statement of Eleanor Hill).

26. U.S. CIA, *KUBARK.*

27. Petula Dvorak, "Cell Phones' Flaws Imperil 911 Response," *Washington Post,* March 31, 2003, sec. B, http://www.washingtonpost.com/ac2/ .

wp-dyn?pagename=article&node=&contentId=A54802-2003Mar30&not Found=true.

28. National Emergency Number Association, "About/Contact NENA," http://www.nena9-1-1.org/About_Contact/index.htm (accessed September 14, 2003).

29. National Emergency Number Association, "NENA Business Alliance Committee Members Announced," news release, September 11, 2000, http://www.nena9-1-1.org/PR_Pubs/PressReleases/Member%20News%20 Release—NBA%209-11-00.PDF.

30. Matthew Fordahl, "Does Your Car Have a 'Black Box' Recorder?" *Chicago Sun-Times,* July 13, 2003.

31. Associated Press, "Stalkers Use GPS to Track Victims," *Wired News,* February 6, 2003, http://www.wired.com/news/wireless/0,1382,57576,00 .html.

32. Julia Scheeres, "Kidnapped? GPS to the Rescue," *Wired News,* January 25, 2002, http://www.wired.com/news/business/0,1367,50004,00.html.

33. Declan McCullagh, "U.S. Could Deny GPS to Taliban," *Wired News,* October 20, 2001, http://www.wired.com/news/conflict/0,2100,47739,00 .html.

Nothing to Fear

1. John Locke, "Second Treatise, Sections 138–140," quoted in *Two Treatises of Government,* ed. Peter Laslett (New York: Mentor Books, New American Library, 1965).

2. Jonathan Elliot, ed., *The Debates in the Several State Conventions on the Adoption of the Federal Constitution,* 1:449–450.

3. Adam Smith, *An Inquiry into the Nature and Wealth of Nations: A Selected Edition,* ed. Kathryn Sutherland (Oxford: Oxford University Press, 1993), 413.

4. James Gordon Meek, "Ashcroft Tour to Plug Terror Bill," *New York Daily News,* August 6, 2003, http://www.nydailynews.com/08-06-2003/news/wn _report/story/106872p-96686c.html.

5. U.S. CIA, *KUBARK.*

6. Robert O'Harrow Jr., "Air Security Network Advances," *Washington Post*, March 1, 2003, sec. E, http://www.washingtonpost.com/ac2/wp-dyn ?pagename=article&node=&contentId=A18601-2003Feb28&notFound =true.

7. O'Harrow, "Air Security."

8. Steve Johnson, "Critics Point to Snoop Factor in Airline Security," *Detroit Free Press,* June 10, 2003, http://www.freep.com/money/business/capps10_20030610.htm.

9. U.S. Department of Transportation, Office of the Secretary, *Privacy Act of 1974: System of Records,* "Notice to amend a system of records," in *Federal Register* 68, no. 10 (January 15, 2003): 2101–2103.

10. According to the ACLU, cited by Michelle Delio, "Privacy Activist Takes on Delta," *Wired,* March 5, 2003, http://www.wired.com/news/privacy/0,1848,57909,00.html.

11. Bruce Schneier, "Terror Profiles By Computers Are Ineffective," *Newsday,* October 21, 2003, http://www.newsday.com/news/opinion/nyvpsch2135 03428oct21,0,3927478.story.

12. In June, the Electronic Privacy Information Center actually did sue the federal government to obtain details on how CAPPS-2 would work, and to uncover the U.S. Defense Department's role in CAPPS, after discovering a memo to the TSA from John Poindexter, the head of the Pentagon's Office of Information Awareness. Steve Johnson, "Suit Seeks Details of Airline Passenger Screening Network," *San Jose Mercury News,* June 13, 2003, http://www.siliconvalley.com/mld/siliconvalley/6079142.htm.

13. Boycott Delta. http://www.boycottdelta.org (accessed November 9, 2003).

14. Michelle Delio, "CAPPSII Testing on Back Burner," *Wired,* June 13, 2003, http://www.wired.com/news/privacy/0,1848,59252,00.html.

15. Sara Kehaulani Goo, "Fliers to Be Rated for Risk Level," *Washington Post,* September 9, 2003, http://story.news.yahoo.com/news?tmpl=story&cid=18 02&ncid=1802&e=2&u=/washpost/20030909/ts_washpost/a45434_2003sep8.

16. Goo, "Fliers."

17. Goo, "Fliers."

18. Goo, "Fliers."

19. Delio, "Privacy Activist."

20. Associated Press, "Airlines' 7 Passenger Privacy Principles," March 17, 2004.

21 U.S. CIA, *KUBARK.*

22. Foucault, *Discipline & Punish,* 201.

23. Christopher Dandeker, *Surveillance, Power and Modernity: Bureaucracy and Discipline from 1700 to the Present Day* (Cambridge, UK: Polity Press, 1990), vii.

24. Foucault, *Discipline & Punish,* 202.

25. Weber, *Economy and Society,* 255.

26. According to Christopher Dandeker, Weber argued that "bureaucratic administration is based on rational discipline . . . [and is] analogous to the machine in the extent to which subjective or irrational elements of will and mood are eliminated. . . . For Weber, then, rational administration is a fusion of knowledge and discipline." Dandeker, *Surveillance, Power and Modernity,* 9–10.

27. B. Traven, *The Death Ship* (Brooklyn, N.Y.: Lawrence Hill, 1991), originally published as *Totenschiff* (Berlin: Buchmeister Verlag, 1926). Quoted in the *Anderson Valley Advertiser,* September 17, 2003, 10.

28. Anatole France, *The Red Lily My Friend's Book* (n.p., 1885).

29. Kaluszynski, "Republican Identity," 123.

30. Jeremy Benthan, "Letter XXI: Schools," *Panopticon Letters,* http://cartome.org/panopticon2.htm (accessed September 9, 2003).

The Real World

1. Scientists are slowly rediscovering the biochemical and neurological processes by which repressed emotions are stored in the body, to come back later as habitual emotional projections and compulsive behavior. The unconscious isn't mental, as Western psychologists thought—it's in the body, waiting to be heard. See Candace B. Pert, *Molecules of Emotion: The Science Behind Mind-Body Medicine* (New York: Touchstone, 1997). See also Andrew Newberg, Eugene D'Aquili, and Vince Rause, *Why God Won't Go Away: Brain Science and the Biology of Belief* (New York: Ballantine Books, 2002).

2. And don't give us any nonsense about stones not having volition simply because they can't walk. It would make just as much sense for rocks to say humans do not have volition because we can't survive underground, or because we can't retain form over millions of years. All beings have volition within their physical capacities. Humans cannot choose to flap our wings and fly, because we do not have wings (and we're not talking about sitting in an uncomfortable chair [not next to an ax murderer, thank goodness] going from Chicago to Kansas City; we're talking about flying). Nor can we choose to stand rooted in place for a thousand years. Nor can we choose to extend roots into the soil at all. Nor can we choose to dive to a thousand feet underwater. You get the picture: we all make choices according to our physical capacities. I sometimes wonder if flowers pity us because we do not get to bring bees (or the wind) into our lovemaking.

3. University of Southern California News Service, "Machine Demonstrates Superhuman Speech Recognition Abilities," news release 0999025,

September 30, 1999, http://www.fas.org/irp/program/process/36013.htm (accessed September 13, 2003).

4. They do, of course, have trouble figuring out how to stop those who are interested in money and power from killing them.

Money

1 Caplan and Torpey, eds., *Documenting Individual Identity,* 1.

2. Lee, "Passports and Visas."

3. CBC News, "No Case Made for ID Cards: Privacy Commissioner," September 19, 2003, http://www.cbc.ca/stories/2003/09/19/idcard030919.

4. Brian Krebs, "Ex-Officials Urge U.S. To Boost Cybersecurity," *Washington Post,* April 9, 2003, sec. E.

5. Bill Berkowitz, "AmeriSnitch," *The Progressive,* May 24, 2002.

6. Angel Paez, "CIA Gave $10 Million to Peru's Ex-Spymaster," *The Public i,* July 3, 2001.

7. Andy Sullivan, "Military Says Computer Dragnet to Include Limits," Reuters, May 20, 2003.

8. Andrea Elliott, "Stores Fight Shoplifting With Private Security," *New York Times,* June 17, 2003.

9. Lori Valigra, "Fabricating the Future," *Christian Science Monitor,* August 29, 2002, http://csmweb2.emcweb.com/2002/0829/p11s01-stgn.html.

10. Bill Berkowitz, "Protestors Are Not Terrorists," WorkingForChange.com, June 16, 2003, http://www.workingforchange.com.

11. Kari Lydersen, "Spying for Fun and Profit," *AlterNet,* May 28, 2003, http://www.alternet.org/story.html?StoryID=16009.

12. *MacNeil/Lehrer Online NewsHour,* "The USA PATRIOT Act," http://www.pbs.org/newshour/bb/terrorism/homeland/patriotact.html (accessed June 30, 2003).

13. Sara Kehaulani Goo, "An ID With a High IQ," *Washington Post,* February 23, 2003, sec. H, http://www.washingtonpost.com/wp-dyn/articles/A45428-2003Feb21.html.

14. Randy Barrett, "Ridge: Centralized Tech Spending Key to Homeland Security," *National Journal's Technology Daily,* June 26, 2003, http://www.govexec.com/dailyfed/0603/062603td2.htm.

15. Tanya N. Ballard, "Homeland IT Costs May Be Understated, GAO Warns," GovExec.com Daily Briefing, January 14, 2003, http://www.govexec.com/dailyfed/0103/011403t2.htm.

16. Michael Kanellos, "Nanotech Spending Nears $3 Billion," *CNET News,* June 25, 2003, http://news.com.com/2100-1008_3-1021129.html; and Tim Radford, "Brave New World or Miniature Menace?" *The Guardian* (UK), April 29, 2003, http://www.guardian.co.uk/uk_news/story/0,3604,945498 ,00.html.

17. ETC Group, *The Big Down,* 5; and *Holmes Report,* "Brodeur Launches New Nanotechnology Practice," June 3, 2003, http://www.holmesreport .com/holmestemp/story.cfm?edit_id=3310&typeid=1.

18. James Maguire, "Who's Spying on You at Work?" *NewsFactor.com,* July 18, 2003, http://story.news.yahoo.com/news?tmpl=story2&u=/nf/20030718/ tc_nf/21925&e=5.

19. Sniffen, "Pentagon Developing System."

20. U.S. Department of Justice, Bureau of Justice Statistics, "Expenditure and Employment Statistics," http://www.ojp.usdoj.gov/bjs/eande.htm (accessed September 22, 2003).

21. Christian Parenti, *Lockdown America: Police and Prisons in the Age of Crisis* (London and New York: Verso Books, 1999), http://www.thirdworld traveler.com/Prison_System/BigBucks_BigHouse_LA.html (accessed September 22, 2003).

22. *Mother Jones,* "Prison Spending Growing Six Times Faster than Higher Education Spending," July 11, 2001, http://www.motherjones.com/about _us/pressroom/prisons_release.html.

23. Vince Beiser, "How We Got To Two Million," *Mother Jones,* July 10, 2001, http://www.motherjones.com/prisons/overview.html.

24. UNICOR, *2002 Annual Report, Statements of Operations and Cumulative Results of Operations,* http://www.unicor.gov/about/2002annual/auditors _report05.htm (accessed September 22, 2003).

25. Anitha Reddy, "U.S. Readies Program to Track Visas," *Washington Post,* September 29, 2003, http://www.washingtonpost.com/wp-dyn/articles/ A14287-2003Sep28.html.

26. Acuity Market Intelligence, http://acuity-mi.com/?page=home_biomet rics/index (accessed September 21, 2003).

27. Warren Brown, "Auto Shows Reveal Technological Gems," *Washington Post,* January 7, 2003, http://www.washingtonpost.com/wp-dyn/articles/ A22287-2003Jan7.html.

28. Elisa Batista, "Step Back for Wireless ID Tech?" *Wired,* April 8, 2003, http://www.wired.com/news/wireless/0,1382,58385-2,00.html.

29. Kayte VanScoy, "Can the Internet Hot-Wire P&G?: They Know What You Eat," *Ziff Davis Smart Business,* January 1, 2001.

30. Charles W. Schmidt, "The Networked Physical World," http://www .rand.org/scitech/stpi/ourfuture/Internet/sec4_networked.html.

31. Brown, "Auto Shows."

32. Chai and Shim, "Benetton Considers Chip Plans."

33. Declan McCullagh, "Are Spy Chips Set to Go Commercial?" *ZDNet,* January 13, 2003, http://zdnet.com.com/2100-1107-980345.html.

34. Andy McCue, "Gillette Shrugs Off RFID-tracking Fears," *New York Times,* August 15, 2003, http://www.nytimes.com/cnet/CNET_2100-1039 _3-5063990.html?ex=1061959699&ei=1&en=e3f11c9e48afdafd.

35. Martin Kady, "Matrics Bets Chips on $14M Launch," *Washington Business Journal,* January 14, 2002, http://washington.bizjournals.com/washington/ stories/2002/01/14/story3.html.

36. Benetton, "No Microchips Present in Garments on Sale," news release, April 4, 2003, http://www.benetton.com/press/sito/_media/press_releases/ rfiding.pdf.

37. Batista, "Step Back for Wireless ID Tech?"

38. Associated Press, "Lawsuits Cite Consumer Racial Profiling," *Yahoo News,* June 9, 2003, http://story.news.yahoo.com/fc?cid=34&tmpl=fc&in =Business&cat=Consumer_News_and_Recall_Information.

39. Chris E. McGoey, "Shoplifting: Racial Profiling," http://www.crimedoctor .com/shoplifting5.htm (accessed August 17, 2003).

40. McGoey, "Shoplifting."

41. VeriChip, "Products and Services. VeriChip," http://www.adsx.com/prod servpart/verichip.html (accessed September 13, 2003).

42. Scheeres, "Kidnapped? GPS to the Rescue."

43. Auto-ID Center, "Identifying Trillions of Items," http://www.autoidcenter .org/aboutthetech_identifying.asp (accessed August 14, 2003).

44. *Budget of the United States Government, Fiscal Year 2003,* "Department of Justice, 2003 Counter Terrorism Enhancement," chap. 18, 204–205. http://www.gpo.gov/usbudget/fy2003/pdf/bud18.pdf.

45. Benjamin Weiser, "F.B.I. Accused of Corrupting Computer Surveillance," *New York Times,* August 20, 2003, http://www.nytimes.com/2003/08/20/ nyregion/20STEW.html.

46. *CBS News,* "Gitmo Interrogations Shaky," April 21, 2002, http://www.cbsnews.com/stories/2002/04/22/attack/main506836.shtml.

47. U.S. Congress, Hearing before the Committee on Appropriations, Subcommittee on the Departments of Commerce, Justice and State, the Judiciary and Related Agencies, February 28, 2002 (testimony of U.S. Attorney General John Ashcroft). http://www.usdoj.gov/ag/testimony/2002/FY2003AG_WrittenStatement-House.htm.

48. Agence France-Presse, "Prison Population Growth Pushes Number of US Inmates to Over 2.1 Million," *Yahoo News,* July 28, 2003, http://asia.news.yahoo.com/030728/afp/030728070616int.html. Let's break down the two million inmates: 1.2 million in state prisons, 151,000 in federal facilities, 665,000 in local jails, 110,000 juveniles in public and private facilities, 8,700 held by immigration and customs services. China, with a population of 1.3 billion people, has 1.4 million of them in prison; Russia, another totalitarian state, has 920,000 prisoners behind bars.

49. U. S. Congress, Hearing before the Committee on Appropriations (testimony of U.S. Attorney General John Ashcroft).

50. Seisint, "Seisint Overview," http://www2.seisint.com/aboutus/index.html (accessed September 24, 2003).

51. Robert O'Harrow Jr., "U.S. Backs Florida's New Counterterrorism Database," *Washington Post,* August 6, 2003, sec. A.

52. *Sun-Sentinel* (Fort Lauderdale, Fla.), "State Law Enforcement Contractor Linked to Drugs," August 3, 2003, http: // mapinc.org/drugnews/v03/n11 71/a06.html.

53. O'Harrow, "U.S. Backs Florida's."

54. *Sun-Sentinel,* "State Law Enforcement Contractor Linked To Drugs."

55. Peter Landesman, "Arms and the Man," *New York Times,* August 17, 2003, http://www.nytimes.com/2003/08/17/magazine/17BOUT.html.

56. L-3 Communications, "L-3 Communications to Participate in $498 Million Contract to Provide Air Force Base Security Systems," *Business Wire,* September 4, 2003, http://search.hoovers.com/free/co/news/detail.xhtml?COID=89493&ArticleID=NR20030904290.2_c23d0016273183b6.

57. Amy Kover, "A Tech Company Wins Big in the War on Terror," *New York Times,* September 14, 2003, http://www.nytimes.com/2003/09/14/technology/14COMM.html.

58. ChoicePoint, "Overview," http://www.choicepoint.net/about/overview.html (accessed August 19, 2003).

59. Derek V. Smith, Chairman and Chief Executive Officer of ChoicePoint, "Remarks at ChoicePoint 2003 Annual Shareholders Meeting," April 29, 2003, http://www.choicepoint.net/news/feature042903.html. ChoicePoint stock has risen in value an average of 30 percent a year for the past five years.

60. ChoicePoint, "Privacy Policy," http://www.choicepoint.net/privacy.html (accessed August 19, 2003).

61. Equifax, "Frequently Asked Questions," https://www.econsumer.equifax .com/consumer/forward.ehtml?forward=fyi (accessed April 19, 2004).

62. *Hoover's,* "The Marmon Group, Inc.," http://www.hoovers.com/free/co/ factsheet.xhtml?COID=40296 (accessed September 13, 2003).

63. Sybase advertisement published in *Forbes,* April 14, 2003, 6–7.

64. Sybase, "About Sybase," http://www.sybase.com/about_sybase (accessed August 19, 2003).

65. The Chatterjee Group, including its Winston Partners, owns 5.5 million shares in Sybase. The Sybase-Winston connection is from Margie Burns, "Bush Family Dipping Into Security Pie," *Prince George's Journal* (Md.), November 27, 2002, http://www.commondreams.org/views02/1127-07.htm.

66. Philip Shenon, "Former Domestic Security Aides Switch to Lobbying," *New York Times,* April 29, 2003, http://www.nytimes.com/2003/04/29/ politics/29HOME.html.

67. *CBS News*/Associated Press, "Top DOJ Official Stepping Down," August 11, 2003, http://www.cbsnews.com/stories/2003/08/11/national/main567 619.shtml.

68. Corporate crime, however defined and measured, is a small fraction of the subsidies that prop up what is called "free-market" capitalism™. The whole economic system is actually based on subsidies, that is, the externalization of costs. Indeed, it's not too much to say that the primary purpose of government is to oversee and administer this process, and to neutralize or kill anyone who too strongly opposes it. The entire economy would collapse immediately without constant massive subsidies of money taken from the public as taxes and then handed over to various sectors of the economy as "incentives." These tax subsidies range from bailing out industries (banks, airlines, auto manufacturers), to tax breaks (most of the largest corporations pay little or no corporate income tax), to the whole military-industrial complex. Tax subsidies cost American taxpayers billions of dollars each year, but these are only the tip of the externalization iceberg. Indirect subsidies are far more onerous. Workplace injuries cost Americans more than $100 billion a year, and workplace cancer costs us

more like $300 billion. Price-fixing and false advertising costs American consumers more than $1 trillion a year. Air pollution causes more than $200 billion a year in health care. But these are only the current monetary damages. See Ralph Estes, *Tyranny of the Bottom Line: Why Corporations Make Good People Do Bad Things* (San Francisco: Berrett-Koehler, 1996), 177–78. The global trading system results in the transfer of wealth from the poor to the wealthy. When industrial civilization destroys the productive capacity of soils and forests, the reduction in productivity and the quality of life are passed on to future generations. The true costs of over-consumption by humans are also paid by other species, with their lives.

69. *CBS*/AP, "Top DOJ Official Stepping Down."

70. Associated Press, "CIA Funds New Ventures With In-Q-Tel," *USA Today*, January 10, 2002, http://www.usatoday.com/tech/techinvestor/2002/01/10/cia-vc-fund.htm; and Rick E. Yannuzzi, "In-Q-Tel: A New Partnership Between the CIA and the Private Sector," *Defense Intelligence Journal* 9, no. 1 (Winter 2000), http://www.cia.gov/cia/publications/inqtel/. In-Q-Tel's Web site is www.in-q-tel.com.

71. Douglas Jehl, "Insiders' New Firm Consults on Iraq," *New York Times*, September 30, 2003, http://www.nytimes.com/2003/09/30/politics/30LOBB.html.

The Noose Tightens

1. European Parliament Temporary Committee on the ECHELON Interception System, "Final Report on the Existence of a Global System for the Interception of Private and Commercial Communications," September 5, 2001, 35, http://www.fas.org/irp/program/process/rapport_echelon_en.pdf.

2. James Bamford, *Body of Secrets: Anatomy of the Ultra-Secret National Security Agency from the Cold War Through the Dawn of a New Century* (New York: Doubleday, 2001), 40, 403–404.

3. European Parliament Temporary Committee, "Final Report."

4. European Parliament Temporary Committee, "Final Report," 31–35, 55–59.

5. European Parliament Temporary Committee, "Final Report," 36.

6. U.S. Federal Bureau of Investigation, "Carnivore: Diagnostic Tool," http://www.fbi.gov/hq/lab/carnivore/carnivore2.htm (accessed October 28, 2003).

7. Russell Mokhiber, *Corporate Crime and Violence* (San Francisco: Sierra Club Books, 1988), 16–17.

8. U.S. FBI, "Carnivore."

9. Omar J. Pahati, "Confounding Carnivore: How to Protect Your Online Privacy," *AlterNet,* November 29, 2001.

10 *MacNeil/Lehrer,* "The USA PATRIOT Act."

11. Council of Europe, "Convention on Cybercrime, Budapest, November 23, 2001," http://conventions.coe.int/Treaty/en/Treaties/html/185.htm. See also analysis at American Civil Liberties Union, "The International Cybercrime Convention: What Is It?" March 17, 2002, http://www.aclu.org/Privacy/Privacy.cfm?ID=9996&c=130.

12. American Civil Liberties Union, "Is the U.S. Turning Into a Surveillance Society?" http://www.aclu.org/Privacy/Privacylist.cfm?c=39 (accessed August 18, 2003).

13. Jay Stanley and Barry Steinhardt, "Bigger Monster, Weaker Chains: The Growth of an American Surveillance Society," American Civil Liberties Union, 2003, http://www.aclu.org/Files/getFile.cfm?id=11572.

14. ACLU, "Is the U.S. Turning?"

15. Stanley and Steinhardt, "Bigger Monster," preface.

16. Stanley and Steinhardt, "Bigger Monster," 14–17.

17. Jerry Berman and Paula Bruening, "Is Privacy Still Possible in the Twenty-first Century?" *Social Research,* Spring 2001, http://www.cdt.org/publications/privacystill.shtml.

18. Gandy, *The Panoptic Sort,* 55.

19. Antonio Machado, "Proverbios y Cantares," from *Times Alone: Selected Poems of Antonio Machado,* trans. and selected by Robert Bly (Middleton, Conn.: Wesleyan University Press, 1983), 143.

20. Reporters Committee for Freedom of the Press, "Homefront Confidential: How the War on Terrorism Affects Access to Information and the Public's Right to Know," 4th ed., September 2003, http://www.rcfp.org/homefrontconfidential/foi.html.

21. Brett Warneke, "Smart Dust," University of California, Berkeley, Department of Electrical Engineering and Computer Sciences, http://www-bsac.eecs.berkeley.edu/~warneke/SmartDust/index.html (accessed August 11, 2003). See also Craig Matsumoto, "Intel Envisions Smart Dust," *Light Reading*, March 20, 2003, http://www.lightreading.com/document.asp?doc_id=30018.

22. Great Duck Island, "Habitat Monitoring on Great Duck Island," http://www.greatduckisland.net (accessed August 11, 2003).

23. Brian McDonough, "Networked Computer Sensors Infiltrate Everything," *NewsFactor Network,* June 5, 2002, http://sci.newsfactor.com/perl/story/18088.html.

24. Unless otherwise cited, the information on UAVs is from NOVA, *Spies That Fly,* Public Broadcasting System (PBS), http://www.pbs.org/wgbh/nova/spiesfly/.

25. Global Security, "RQ-1 Predator MAE UAV MQ-9A Predator B," http://www.globalsecurity.org/intell/systems/predator.htm (accessed October 23, 2003).

26. Duncan Campbell, "US Buys Up All Satellite War Images," *The Guardian* (UK), October 17, 2001, http://www.guardian.co.uk/Archive/Article/0,4273,4278871,00.html.

27. Vernon Loeb, "Concrete-Piercing Bombs Hammer Caves," *Washington Post,* December 13, 2001, http://www.globalsecurity.org/org/news/2001/011213-attack02.htm; and Shyam Bhatia, "US Forces Use Daisy Cutters," *Rediff.com,* April 3, 2003, http://www.rediff.com/us/2003/apr/03iraq6.htm.

28. Jim Krane, "Pentagon Could Debut New Weapons in Iraq," Associated Press, February 18, 2003.

29. Jonathan S. Landay, "Bold U.S. Strike in Yemen Kills 6," *Mercury News* (San Jose, California), November 5, 2002, http://www.bayarea.com/mld/mercurynews/news/4446720.htm; Craig Pincus, "U.S. Strike Kills Six in Al Qaeda," *Washington Post,* November 5, 2002, http://www.washingtonpost.com/ac2/wp-dyn/A5126-2002Nov4; Craig Hoyle, "Yemen Drone Strike: Just the Start?" *Jane's Defence Weekly,* November 8, 2002, http://www.janes.com/aerospace/military/news/jdw/jdw021108_1_n.shtml; Dana Priest, "U.S. Citizen Among Those Killed In Yemen Predator Missile Strike," *Washington Post,* November 8, 2002; and *ABC News.com,* "Predator Drone Kills Six Al Qaeda Suspects," October 21, 2002, http://abcnews.go.com/sections/wnt/DailyNews/yemen021105.html.

30. *World Tribune,* "In Wake of Predator Success, U.S. Weighs Assassination Options," November 29, 2002, http://216.26.163.62/2002/ss_terrorism_11_29.html. The U.S. government had indicted one of the six men killed by a Predator drone's missile in Yemen for being involved in the October 2000 attack on a U.S. Navy's destroyer. For an article about the legality of such killings, see Nyier Abdou, "Death by Predator," *Al-Ahram Weekly Online,* November 20, 2002, http://weekly.ahram.org.eg/2002/612/re5.htm.

31. Global Security, "RQ-1 Predator."

32. Al Martin, "Coming Soon: Flying Fascism on Your Doorstep," *Pravda,*

February 20, 2002, http://english.pravda.ru/columnists/2002/02/20/265 31.html.

33. Charles Piller, "Army of Extreme Thinkers," *Los Angeles Times,* August 14, 2003, http://www.latimes.com/news/science/la-sci-darpa14aug14,1,61 43132.story?coll=la-news-science.

34. Andrew Pollack, "Army Center to Study New Uses of Biotechnology," *New York Times,* August 27, 2003, http://www.nytimes.com/2003/08/27/ national/27BIOT.html.

35. U.S. National Institute of Standards and Technology, "Technologies for Improved Homeland Security," http://www.nist.gov/public_affairs/ factsheet/homeland.htm (accessed September 8, 2003).

36. Biometric Consortium, "Background of the U. S. Government's Biometric Consortium," http://www.biometrics.org/REPORTS/CTST96/ (accessed August 7, 2003).

37. *BioMetriTech,* "Precise Receives Order From Stockholm City Schools," August 6, 2003, http://www.biometritech.com/enews/080603a.htm.

38. U.S. National Institute of Standards and Technology, "Biometric Consortium Conference, September 23–25, 2002. About Biometrics," http://www.itl.nist.gov/div895/isis/bc/bc2002/aboutbiometrics.htm (accessed September 13, 2003).

39. U.S. National Institute of Standards and Technology, "General Information," http://www.nist.gov/public_affairs/general2.htm (accessed September 13, 2003).

40. U.S. National Institute of Standards and Technology, "NIST In Your House," http://www.nist.gov/public_affairs/nhouse/index.html (accessed September 13, 2003).

41. U.S. National Institute of Standards and Technology, "Biometric Interoperability, Performance, and Assurance Working Group," http://www.itl.nist.gov/div895/isis/bc/bcwg/ (accessed September 13, 2003).

42. Michelle Shen, "The 'People' Element In Biometrics And Physical Access Control," *BioMetriTech,* April 14, 2003, http://www.biometritech.com/ features/shen041403.htm.

43. Laura Guevin, "Striking While The Iron Is Hot," *BioMetriTech,* April 29, 2003, http://www.biometritech.com/features/notebook042903.htm.

44. Michelle Shen, "Trends In Biometric Security (Part 3): Buyer Behavior Analysis," *BioMetriTech,* March 19, 2003, http://www.biometritech.com/ features/shen031903.htm.

45. U.S. Department of Defense, Biometrics Management Office, "Frequently

Asked Questions," https://www.bfc-kno.army.mil/faq/faq.htm (accessed August 7, 2003).

46. Brain Fingerprinting Laboratories, "About Brain Fingerprinting Laboratories," http://www.brainfingerprinting.com/about-bfl.php (accessed June 18, 2003).

47. Brain Fingerprinting Laboratories, "Counterterrorism Applications," http://www.brainfingerprinting.com/counterterrorism.php (accessed June 18, 2003).

48. Brain Fingerprinting Laboratories, "Scientific Procedure, Research, and Applications," http://www.brainfingerprinting.com/TechnologyOver view.php (accessed June 18, 2003).

49. Brain Fingerprinting Laboratories, "Scientific Procedure."

50. U.S. Senator Charles Grassley, quoted at http://www.brainfingerprinting .com/HomePage.php (accessed June 18, 2003).

The End

1. U.S. Department of Defense, "Assistant Secretary of Defense for Networks and Information Integration/Department of Defense Chief Information Officer," http://www.dod.mil/nii/bio/asd/ (accessed September 13, 2003).

2. U.S. DARPA, "Legacy," http://www.darpa.mil/body/legacy/prev _items.html (accessed August 18, 2003).

3. Bob Drogin and Aaron Zitner, "No Drivers Wanted in Race for $1 Million," *Los Angeles Times,* February 21, 2003, sec. A, http://www.la times.com/news/nationworld/nation/la-na-race21feb21,1,5698431.story.

4. U.S. DARPA, "Ability to Simulate Gunshot Wounds Provides Realistic Training," March 27, 2001, http://www.darpa.mil/body/legacy/prev _items.html.

5. U.S. DARPA, "BioSPICE Project Proposal Solicitation," http://www .darpa.mil/ito/Solicitations/PIP_01-26.html (accessed July 1, 2003).

6. Two articles in the same day's *New York Times* illustrate the point: for the police that look like soldiers, see Shaila K. Dewan, "Cargo Pants, and They Come With Cuffs," *New York Times,* August 26, 2003, http://www .nytimes.com/2003/08/26/nyregion/26CARG.html. For the rise of warrior fashion, see Ginia Bellafante, "Suiting Up With the New Woman Warrior," *New York Times,* August 26, 2003, http://www.nytimes.com/ 2003/08/26/fashion/26DRES.html.

7. Alexander Gourevitch, "Body Count: How John Ashcroft's Inflated

Terrorism Statistics Undermine the War on Terrorism," *Washington Monthly*, June 2003, http://www.washingtonmonthly.com/features/2003/0306.gourevitch.html.

8. Lewis Mumford, *The Myth of the Machine, Volume I: Technics and Human Development* (New York: Harvest/HBJ, 1966), 213.

9. Elan Journo, "Under Attack—by Eco-Terrorists: Homeland Security, Stopping the Destruction of Property in Defense of Nature," *San Francisco Chronicle*, October 12, 2003, sec. D, http://www.sfgate.com/cgi-bin/article.cgi?file=/chronicle/archive/2003/10/12/INGET28F3B1.DTL.

10. Lewis Mumford, *The City in History: Its Origins, Its Transformations, and Its Prospects* (New York: Harbinger, 1961), 60.

11. Mumford, *The Myth of the Machine, Volume I*, 12.

12. Tim Haynes, "High-tech Robot Gives Glimpse of Future," *Boston Globe*, July 25, 2003, http://www.boston.com/dailyglobe2/206/metro/High_tech_robot_gives_glimpse_of_future+.shtml.

13. Reuters, "Robot Shows Prime Minister How to Loosen Up," August 22, 2003.

14. United Press International, *Android World*, August 2, 2003, http://www.androidworld.com/ (accessed August 11, 2003).

15. U.S. DARPA, "Autonomous Vehicles Grand Challenge," http://www.darpa.mil/grandchallenge/overview.htm (accessed August 11, 2003).

16. Witold Rybcynski, *Taming the Tiger: The Struggle to Control Technology* (New York: Viking Press, 1983), vii, 227.

17. Jean-Jacques Rousseau, *On the Social Contract* (Mineola, New York: Dover Publications, 2003), 61. Had Rousseau been alive in 2003, he may very well have added, ". . . and even become President of the United States."

18. Rousseau, *On the Social Contract*, 3.

19. Éttiene de La Boétie, *Discours de la Servitude Volontaire* (1552), trans. Harry Kurz, *The Politics of Obedience: The Discourse of Voluntary Servitude* (Buffalo, New York: Black Rose Books, 1997). Our quote comes from the abridged and edited version at Personal Empowerment Resources, http://www.mind-trek.com/ (accessed November 1, 2003).

20. Pilar Viladas, "Home Despot," *New York Times Magazine*, August 17, 2003, http://www.nytimes.com/2003/08/17/magazine/magazinespecial/WFOTHOMET.html.

21. Joseph Menn, "Techies, Politics Now Click," *Los Angeles Times*, August 11, 2003, http://www.latimes.com/business/la-fi-geeks11aug11000416,1,3687057.story?coll=la-headlines-business-manual.

22. World Summit on the Information Society, "Basic Information: About WSIS," http://www.itu.int/wsis/basic/about.html (accessed November 9, 2003).

23. World Summit on the Information Society, "Information on Business Input," http://www.iccwbo.org/home/e_business/wsis.asp (accessed November 9, 2003).

24. World Summit on the Information Society, "Information on Business Input."

25. World Summit on the Information Society, "Information on Business Input."

26. ICC's Web site is www.iccwbo.org. For profiles of the ICC, the WEF, and other international corporate institutions, see George Draffan, *The Elite Consensus: When Corporations Wield the Constitution* (New York: Apex Press and the Program on Corporations, Law & Democracy, 2003).

27. Editorial, "What Is Operation TIPS?" *Washington Post,* July 14, 2002, http://www.washingtonpost.com/ac2/wp-dyn?pagename=article&node =&contentId=A63924-2002Jul12&notFound=true.

28. Hannah Arendt, *Eichmann in Jerusalem: A Report on the Banality of Evil* (New York: Viking, 1964), 276.

29. Barbarism is precisely the wrong word here. In fact, Merton is talking about the insanity of the civilized. *Barbarians* is a pejorative word for the indigenous, and comes from the ancient Greek belief that because the indigenous cultures they encountered did not speak Greek (the one true language, as civilization was the one true way to be), they did not speak at all, but merely said bah-bah-bah. It is the civilized, not the barbarians, who have committed the worst atrocities.

30. Thomas Merton, "A Devout Meditation on Adolph Eichmann," in *Raids on the Unspeakable* (New York: New Directions, 1966), 45–46.

31. George M. Kren and Leon Rappoport, *The Holocaust and the Crisis of Human Behavior* (New York: Holmes & Meier, 1980), 64.

32. Kren and Rappoport, *The Holocaust,* 70.

33. James H. McRandle, *The Track of the Wolf: Essays on National Socialism and its Leader, Adolf Hitler* (Evanston, Ill.: Northwestern University Press, 1965), 125.

Humanity

1. H. Gillespie and J. Coots, *Santa Claus Is Coming to Town,* lyrics.

2. Orwell, *1984,* 210–211.

3. James Scott, *Weapons of the Weak: Everyday Forms of Peasant Resistance* (New Haven, Conn.: Yale University Press, 1985), 29.

4. John Gilliom, *Overseers of the Poor: Surveillance, Resistance, and the Limits of Privacy* (Chicago: University of Chicago Press, 2001), 102.

5. James Scott, *Domination and the Arts of Resistance: Hidden Transcripts* (New Haven, Conn.: Yale University Press, 1990), chap. 7.

6. Gilliom, *Overseers of the Poor,* 102.

7. Klaus Lenk, "Information Technology and Society," in *Microelectronics and Society: For Better or Worse,* ed. Gunter Friedrichs and A. Schaff (Oxford, UK: Pergamon Press, 1982), 284.

8. "Examining the pursuit of power . . . and [the] balances between technology, armed force, and society will not solve contemporary dilemmas. It may, nonetheless, provide perspective and . . . awareness [and] make simple solutions and radical despair both seem less compelling." William H. McNeill, *The Pursuit of Power: Technology, Armed Force, and Society since A.D. 1000* (Oxford, UK: Basil Blackwell, 1983), viii.

9. Senate Select Committee, *U.S. Congress, Senate Select Committee to Study Governmental Operations with Respect to Intelligence Activities, Final Report,* 94th Cong., 2d sess., 1976, S. Rep. 94-755. The text of Volume 2, *Intelligence Activities and the Rights of Americans*, which includes the committee's recommendations, can be found at http://www.icdc.com/ ~paulwolf/cointelpro/churchfinalreportIId.htm.

10. Lao-Tzu, *Taoteching,* trans. Red Pine (San Francisco: Mercury House, 1996). See especially verses 32, 38, 57, and 80. But take the time to savor the whole poem.

11. Ralph Waldo Emerson, *Works and Days* (1857).

12. J. William Fulbright, speech to the U.S. Senate, April 21, 1966.

13. Vince Beiser, "A Necessary Evil?" *Los Angeles Times,* October 19, 2003, http://www.latimes.com/features/printedition/magazine/la-tm-pelican 42oct19,1,7855478.story?coll=la-headlines-magazine.

14. Charles A. Eastman (Ohiyesa), *Indian Heroes and Great Chieftains* (Boston: Little, Brown, 1918), 14–15, quoted in Bob Blaisdell, ed., *Great Speeches by Native Americans* (Mineola, New York: Dover Thrift Editions, 2000). Red Cloud's words as remembered by George Eastman (Ohiyesa).

15. Bauman, *Modernity and the Holocaust,* 203.

16. La Boétie, *Discours.*

17. Bauman, *Modernity and the Holocaust,* 203.

18. I was told that one can destroy RFID chips by putting them in a microwave oven. So it seemed that the feds and corporations could be foiled just by popping every new sweater and so on into the microwave. I thought that was a good idea until George reminded me what happens when you put metal into a microwave. So we'll have to come up with another way.

Bibliography

Abbey, Edward. Letter to the editor. *Bloomsbury Review,* April 1985.

ABC News.com. "Predator Drone Kills Six Al Qaeda Suspects." October 21, 2002. http://abcnews.go.com/sections/wnt/DailyNews/yemen021105.html.

Abdou, Nyier. "Death by Predator." *Al-Ahram Weekly Online,* November 20, 2002. http://weekly.ahram.org.eg/2002/612/re5.htm.

Act of November 10, 1919, Public Law 79, 66th Cong., sess. 1, ch.104 (November 10, 1919).

Actualizations. "Obsessive Compulsive Disorder." http://www.actualizations .com/ocd.htm#what%20is (accessed October 27, 2003).

Acuity Market Intelligence. http://acuity-mi.com/?page=home_biometrics/ index (accessed September 21, 2003).

Agar, Jon. "Modern Horrors." In *Documenting Individual Identity: The Development of State Practices in the Modern World,* edited by Jane Caplan and John Torpey. Princeton, N.J.: Princeton University Press, 2001.

Agence France-Presse. "Prison Population Growth Pushes Number of US Inmates to Over 2.1 Million." *Yahoo News,* July 28, 2003. http://asia.news.yahoo.com/030728/afp/030728070616int.html.

Albrecht, Katherine. "Auto-ID, Tracking Everything, Everywhere." From "Supermarket Cards: Tip of the Retail Surveillance Iceberg." *Denver University Law Review* 79, no. 4 (Summer 2002).

Alvares, Claude. "Western Science and Violence." Istanbul.indymedia.org. http://istanbul.indymedia.org/news/2003/10/2795.php (accessed October 24, 2003).

American Civil Liberties Union. "The International Cybercrime Convention: What Is It?" March 17, 2002. http://www.aclu.org/Privacy/Privacy.cfm ?ID=9996&c=130.

———. "Is the U.S. Turning Into a Surveillance Society?" http://www.aclu .org/Privacy/Privacylist.cfm?c=39 (accessed August 18, 2003).

American–Israeli Cooperative Enterprise. "Jewish Virtual Library." http://www.us-israel.org/jsource/Holocaust/Hollerith.html (accessed August 16, 2003). Citing United States Holocaust Memorial Museum, http://www.ushmm.org/.

Arendt, Hannah. *Eichmann in Jerusalem: A Report on the Banality of Evil*. New York: Penguin, 1994.

Associated Press. "Airlines' 7 Passenger Privacy Principles." March 17, 2004.

———. "CIA Funds New Ventures With In-Q-Tel." *USA Today,* January 10, 2002. http://www.usatoday.com/tech/techinvestor/2002/01/10/cia-vc-fund .htm.

————. "Lawsuits Cite Consumer Racial Profiling." *Yahoo News,* June 9, 2003. http://story.news.yahoo.com/fc?cid=34&tmpl=fc&in=Business&cat =Consumer_News_and_Recall_Information.

————. "Pentagon Hopes to Identify People by the Way They Walk." *News-Star* (Shawnee, Okla.), May 20, 2003. http://www.news-star.com/stories/ 052003/New_8.shtml.

————. "Stalkers Use GPS to Track Victims." *Wired,* February 6, 2003. http://www.wired.com/news/wireless/0,1382,57576,00.html.

Auto-ID Center. "Identifying Trillions of Items." http://www.autoidcenter .org/aboutthetech_identifying.asp (accessed August 14, 2003).

————. "The Future of Business." http://www.autoidcenter.org/ aboutthecenter.asp (accessed August 14, 2003).

————. "Sponsor Companies." http://www.autoidcenter.org/sponsors _companies.asp (accessed November 5, 2003).

Baard, Erik. "The Guilt-Free Soldier: New Science Raises the Specter of a World Without Regret." *The Village Voice,* January 22–28, 2003. http://www.village voice.com/issues/0304/baard.php (accessed September 2, 2003).

Bacon, Francis. *The Works.* Edited by James Spedding. New York: Garrett Press, 1968.

Ballard, Tanya N. "Homeland IT Costs May Be Understated, GAO Warns." GovExec.com Daily Briefing, January 14, 2003. http://www.govexec.com/ dailyfed/0103/011403t2.htm.

Bamford, James. *Body of Secrets: Anatomy of the Ultra-Secret National Security Agency from the Cold War Through the Dawn of a New Century.* New York: Doubleday, 2001.

Barrett, Randy. "Ridge: Centralized Tech Spending Key to Homeland Security." *National Journal's Technology Daily,* June 26, 2003. http://www.govexec.com/dailyfed/0603/062603td2.htm.

Batista, Elisa. "Step Back for Wireless ID Tech?" *Wired*, April 8, 2003. http://www.wired.com/news/wireless/0,1382,58385-2,00.html.

Bauman, Zygmunt. *Modernity and the Holocaust.* Ithaca, N.Y.: Cornell University Press, 1992.

BBC News. "Health: Breast Milk Studied for Toxins." *BBC Online Network,* July 12, 1999. http://news.bbc.co.uk/1/hi/health/391514.stm (accessed November 5, 2003).

Beiser, Vince. "How We Got To Two Million." *Mother Jones,* July 10, 2001. http://www.motherjones.com/prisons/overview.html.

————. "A Necessary Evil?" *Los Angeles Times,* October 19, 2003. http://www.latimes.com/features/printedition/magazine/ la-tm-pelican42oct19,1,7855478.story?coll=la-headlines-magazine.

Bellafante, Ginia. "Suiting Up With the New Woman Warrior." *New York Times,* August 26, 2003. http://www.nytimes.com/2003/08/26/fashion/26 DRES.html.

Benetton. "No Microchips Present in Garments On Sale." News release, April 4, 2003. http://www.benetton.com/press/sito/_media/press_releases/rfiding.pdf.

Beniger, James. *The Control Revolution: Technological and Economic Origins of the Information Society.* Cambridge, Mass.: Harvard University Press, 1986.

Bentham, Jeremy. *Panopticon Letters.* http://cartome.org/panopticon2.htm (accessed September 9, 2003).

————. *Panopticon; Or The Inspection-House: Containing The Idea Of A New Principle Of Construction Applicable To Any Sort Of Establishment, In Which Persons Of Any Description Are To Be Kept Under Inspection; And In Particular To Penitentiary-Houses, Prisons, Houses Of Industry, Work-Houses, Poor-Houses, Lazarettos, Manufactories, Hospitals, Mad-Houses, And Schools: With A Plan Of Management.* http://cartome.org/panopticon2.htm (accessed June 10, 2003).

————. "Principles of Penal Law." In *The Works of Jeremy Bentham,* vol. 1, edited by John Bowring. Edinburgh: William Tait, 1843.

Berkowitz, Bill. "AmeriSnitch." *The Progressive,* May 24, 2002.

————. "Protestors Are Not Terrorists." WorkingForChange.com, June 16, 2003. http://www.workingforchange.com.

Berman, Jerry, and Paula Bruening. "Is Privacy Still Possible in the Twenty-first Century?" *Social Research* (Spring 2001). http://www.cdt.org/publica tions/privacystill.shtml.

Betterhumans. "Immortality." November 20, 2002. http://www.betterhumans .com/Resources/Goals/goal.aspx?articleID=2002-05-08-5 (accessed September 8, 2003).

Bhatia, Shyam. "US Forces Use Daisy Cutters." *Rediff.com,* April 3, 2003. http://www.rediff.com/us/2003/apr/03iraq6.htm.

Biometric Consortium. "Background of the U.S. Government's Biometric Consortium." http://www.biometrics.org/REPORTS/CTST96/ (accessed August 7, 2003).

BioMetriTech. "Bioscrypt Secures Los Angeles City Hall." July 28, 2002. http://www.biometritech.com/enews/072803b.htm.

————. "Precise Receives Order From Stockholm City Schools." August 6, 2003. http://www.biometritech.com/enews/080603a.htm.

Blaisdell, Bob, ed. *Great Speeches by Native Americans.* Mineola, N.Y.: Dover Thrift Editions, 2000.

Bolonkin, Alexander. "Science, Soul, Heaven and Supreme Mind: Russian Scientist Alexander Bolonkin Develops Artificial Intelligence in the USA."

Pravda, January 8, 2003. http://english.pravda.ru/main/2003/01/08/41749.html.

————. "Twenty-First Century—The Beginning of Human Immortality." http://bolonkin.narod.ru/Bolonkin-p3.htm (accessed September 9, 2003).

Boycott Delta. http://www.boycottdelta.org (accessed November 9, 2003).

Brain Fingerprinting Laboratories. "About Brain Fingerprinting Laboratories." http://www.brainfingerprinting.com/about-bfl.php (accessed June 18, 2003).

————. "Counterterrorism Applications." http://www.brainfingerprinting .com/counterterrorism.php (accessed June 18, 2003).

————. "Scientific Procedure, Research, and Applications." http://www.brain fingerprinting.com/TechnologyOverview.php (accessed June 18, 2003).

Brown, Warren. "Auto Shows Reveal Technological Gems." *Washington Post,* January 7, 2003. http://www.washingtonpost.com/wp-dyn/articles/ A22287-2003Jan7.html.

Browning, Christopher R. *Fateful Months: Essays on the Emergence of the Final Solution.* New York: Holmes & Meier, 1985.

Budget of the United States Government, Fiscal Year 2003. "Department of Justice, 2003 Counter Terrorism Enhancement," chap. 18. http://www.gpo.gov/us budget/fy2003/pdf/bud18.pdf.

Bulkeley, Deborah. "Webcam Surveillance in Biloxi Classrooms: Cameras Watch All Students in School District." *San Francisco Chronicle,* August 13, 2003.

Burns, Margie. "Bush Family Dipping Into Security Pie." *Prince George's Journal* (Md.), November 27, 2002. http://www.commondreams.org/ views02/1127-07.htm.

Campbell, Duncan. "US Buys Up All Satellite War Images." *The Guardian* (UK), October 17, 2001. http://www.guardian.co.uk/Archive/Article/0,42 73,4278871,00.html.

Caplan, Jane. "This or That Particular Person." In *Documenting Individual Identity: The Development of State Practices in the Modern World,* edited by Jane Caplan and John Torpey. Princeton, N.J.: Princeton University Press, 2001.

Caplan, Jane, and John Torpey, editors. *Documenting Individual Identity: The Development of State Practices in the Modern World.* Princeton, N.J.: Princeton University Press, 2001.

Cawthorne, Andrew. "U.S. Raid Herds Iraqi Old and Young in Barbed Wire." *Excite News,* September 1, 2003. http://news.excite.com/world/ article/id/51496|world|09-01-2003::01:48|reuters.html (accessed September 2, 2003).

CBC News. "No Case Made for ID Cards: Privacy Commissioner." September 19, 2003. http://www.cbc.ca/stories/2003/09/19/idcard030919.

CBS News. "Gitmo Interrogations Shaky." April 21, 2002. http://www.cbs news.com/stories/2002/04/22/attack/main506836.shtml.

CBS News/Associated Press. "Top DOJ Official Stepping Down." August 11, 2003. http://www.cbsnews.com/stories/2003/08/11/national/main567619.shtml.

Chai, Winston, and Richard Shim. "Benetton Considers Chip Plans." *CNET News,* April 7, 2003. http://news.zdnet.co.uk/hardware/chips/0,39020354 ,2133031,00.htm.

ChoicePoint. "Overview." http://www.choicepoint.net/about/overview.html (accessed August 19, 2003).

———. "Privacy Policy." http://www.choicepoint.net/privacy.html (accessed August 19, 2003).

CNN News. "Police, School District Defend Drug Raid." *CNN.com,* November 7, 2003. http://www.cnn.com/2003/US/South/11/07/school.raid.

Connecticut Department of Social Services. "DSS's Biometric ID Project: Connecticut Survey of Clients Who Have Just Been Digitally Imaged for AFDC 3/29/96 to 4/3/96." http://www.dss.state.ct.us/digital/faq/ disurvey.htm (accessed August 16, 2003).

———. "DSS's Biometric ID Project: Project Overview." http://www.dss .state.ct.us/digital/project.htm (accessed August 16, 2003).

———. "DSS's Biometric ID Project: Understanding Public Perception." http://www.dss.state.ct.us/Digital/faq/disuppt.htm (accessed August 16, 2003).

Council of Europe. "Convention on Cybercrime, Budapest, November 23, 2001." http://conventions.coe.int/Treaty/en/Treaties/html/185.htm.

Crichton, Michael. *Prey: A Novel.* San Francisco: HarperCollins, 2002.

Dandeker, Christopher. *Surveillance, Power and Modernity: Bureaucracy and Discipline from 1700 to the Present Day.* Cambridge, UK: Polity Press, 1990.

Delio, Michelle. "CAPPSII Testing on Back Burner." *Wired,* June 13, 2003. http://www.wired.com/news/privacy/0,1848,59252,00.html.

———. "Privacy Activist Takes on Delta." *Wired,* March 5, 2003. http://www .wired.com/news/privacy/0,1848,57909,00.html.

Deloria, Vine. "Where The Buffalo Go: How Science Ignores the Living World: An Interview with Derrick Jensen." *The Sun,* July 2000.

Dewan, Shaila K. "Cargo Pants, and They Come With Cuffs." *New York Times,* August 26, 2003. http://www.nytimes.com/2003/08/26/nyregion/ 26CARG.html.

Donnelly, Thomas. "Rebuilding America's Defenses: Strategy, Forces and Resources." The Project for the New American Century, September 2000. http://www.newamericancentury.org/RebuildingAmericasDefenses.pdf.

Draffan, George. *The Elite Consensus: When Corporations Wield the Constitution*. New York: Apex Press and the Program on Corporations, Law & Democracy, 2003.

Drexler, K. Eric. "Engines of Destruction." In *Engines of Creation: The Coming Era of Nanotechnology*. Foresight Institute. http://www.foresight.org/EOC/EOC_Chapter_11.html (accessed September 10, 2003).

———. "Foresight Background 3, Rev. 1: Dialog, Exploratory Engineering, Bioarchive." Foresight Institute. http://www.foresight.org/Updates/Background3.html#DangerDialog (accessed September 10, 2003).

Drogin, Bob, and Aaron Zitner. "No Drivers Wanted in Race for $1 Million." *Los Angeles Times,* February 21, 2003, sec. A. http://www.latimes.com/news/nationworld/nation/la-na-race21feb21,1,5698431.story.

Dvorak, Petula. "Cell Phones' Flaws Imperil 911 Response." *Washington Post,* March 31, 2003, sec. B. http://www.washingtonpost.com/ac2/wp-dyn?pagename=article&node=&contentId=A54802-2003Mar30&notFound=true.

Eastman, Charles A. (Ohiyesa). *Indian Heroes and Great Chieftains*. Boston: Little, Brown, 1918. Reprint edition, Mineola, N.Y.: Dover, 1997.

Elliot, Jonathan, editor. *The Debates in the Several State Conventions on the Adoption of the Federal Constitution*. 5 vols.

Elliott, Andrea. "Stores Fight Shoplifting With Private Security." *New York Times,* June 17, 2003.

Elwell, Frank. "The Sociology of Max Weber." http://www.faculty.rsu.edu/~felwell/Theorists/Weber/Whome.htm (accessed September 16, 2003).

Emerson, Ralph Waldo. *Works and Days*. 1857.

Equifax. "Frequently Asked Questions." https://www.econsumer.equifax.com/consumer/forward.ehtml?forward=fyi (accesed April 19, 2004).

Estes, Ralph. *Tyranny of the Bottom Line: Why Corporations Make Good People Do Bad Things*. San Francisco: Berrett-Koehler, 1996.

ETC Group. *The Big Down: Atomtech: Technologies Converging at the Nanoscale*. Winnipeg, Manitoba: ETC Group, 2003. http://www.etcgroup.org/documents/TheBigDown.pdf.

European Parliament Temporary Committee on the ECHELON Interception System. "Final Report on the Existence of a Global System for the Interception of Private and Commercial Communications," September 5, 2001. http://www.fas.org/irp/program/process/rapport_echelon_en.pdf.

Eveld, Edward M. "It's Time to Break Those Bad Habits: Here's How." *Kansas City Star,* December 29, 2000. http://www.kcstar.com/item/pages/story,local/37750489.c29.html.

Flam, Faye. "Your Brain May Soon Be Used Against You." *Philadelphia Inquirer,*

October 29, 2002. http://www.philly.com/mld/inquirer/4391614.htm.

Fordahl, Matthew. "Does Your Car Have a 'Black Box' Recorder?" *Chicago Sun-Times,* July 13, 2003.

Foucault, Michel. *Discipline & Punish: The Birth of the Prison.* Translated by Alan Sheridan. New York: Vintage Books, 1979.

Fromm, Erich. *The Anatomy of Human Destructiveness.* New York: Holt Rinehart and Winston, 1973.

Fukuyama, Francis. *The End of History and the Last Man.* New York: Free Press, 1992.

Fulbright, J. William. Speech to the U.S. Senate, April 21, 1966.

Galeano, Eduardo. *Upside Down: A Primer for the Looking-Glass World.* Translated by Mark Fried. New York: Picador USA, 2001.

Gandy Jr., Oscar H. *The Panoptic Sort: A Political Economy of Personal Information.* Boulder, Colo.: Westview Press, 1993.

Garcelon, Marc. "Colonizing the Subject." In *Documenting Individual Identity: The Development of State Practices in the Modern World,* edited by Jane Caplan and John Torpey. Princeton, N.J.: Princeton University Press, 2001.

Gartner, John. "Point 'n' Shoot Sound Makes Waves." *Wired News,* February 21, 2002. http://www.wired.com/news/technology/0,1282,50483,00.html (accessed May 20, 2003).

Gillespie, H., and J. Coots. "Santa Claus Is Coming to Town." Lyrics.

Gilliom, John. *Overseers of the Poor: Surveillance, Resistance, and the Limits of Privacy.* Chicago: University of Chicago Press, 2001.

Global Security. "RQ-1 Predator MAE UAV MQ-9A Predator B." http://www.globalsecurity.org/intell/systems/predator.htm (accessed October 23, 2003).

Goldberg, Carey. "Some Fear Loss of Privacy as Science Pries into Brain." *Boston Globe,* May 1, 2003, sec. A.

Goo, Sara Kehaulani. "Fliers to Be Rated for Risk Level." *Washington Post,* September 9, 2003. http://story.news.yahoo.com/news?tmpl=story&cid=1802 &ncid=1802&e=2&u=/washpost/20030909/ts_washpost/a45434_2003sep8.

———. "An ID With a High IQ." *Washington Post,* February 23, 2003, H1. http://www.washingtonpost.com/wp-dyn/articles/A45428-2003Feb21.html.

Gourevitch, Alexander. "Body Count: How John Ashcroft's Inflated Terrorism Statistics Undermine the War on Terrorism." *Washington Monthly,* June 2003. http://www.washingtonmonthly.com/features/2003/0306.gourevitch.html.

Great Duck Island. "Habitat Monitoring on Great Duck Island." http://www.greatduckisland.net (accessed August 11, 2003).

Gruen, Arno. *The Betrayal of the Self: The Fear of Autonomy in Men and Women*. Translated by Hunter and Hildegarde Hannum. New York: Grove Press, 1988.

———. *The Insanity of Normality: Realism as Sickness: Toward Understanding Human Destructiveness*. Translated by Hildegarde and Hunter Hannum. New York: Grove Weidenfeld, 1992.

Guevin, Laura. "Striking While The Iron Is Hot." *BioMetriTech,* April 29, 2003. http://www.biometritech.com/features/notebook042903.htm.

Gugliotta, Guy. "Rats Turned Into Remote-Controlled Robots: Techniques Potential Uses Include Aid to Victims of Disaster or Neural Injuries." *Washington Post,* May 2, 2002, sec. A. http://www.washingtonpost.com/ac2/wp-dyn?pagename=article&node=&contentId=A18261-2002May1&notFound=true.

Haynes, Tim. "High-tech Robot Gives Glimpse of Future." *Boston Globe,* July 25, 2003. http://www.boston.com/dailyglobe2/206/metro/High_tech_robot_gives_glimpse_of_future+.shtml.

Hearst, David. "Sci-fi War Put Under the Microscope." *The Guardian,* May 20, 2003. http://www.guardian.co.uk/uk_news/story/0,3604,959469,00.html (accessed June 21, 2003).

Hilberg, Raul. *Destruction of the European Jews*. Revised and definitive edition. New York: Holmes & Meier, 1985.

Hodge, Rodger D. Weekly Review. *Harper's Magazine,* June 3, 2003.

Holmes Report. "Brodeur Launches New Nanotechnology Practice." June 3, 2003. http://www.holmesreport.com/holmestemp/story.cfm?edit_id=3310&typeid=1.

Hoover's. "The Marmon Group, Inc." http://www.hoovers.com/free/co/factsheet.xhtml?COID=40296 (accessed September 13, 2003).

Hoyle, Craig. "Yemen Drone Strike: Just the Start?" *Jane's Defence Weekly,* November 8, 2002. http://www.janes.com/aerospace/military/news/jdw/jdw021108_1_n.shtml.

Hudson, Audrey. "Tiny IDs Can Track Almost Anything." *Washington Times,* June 9, 2003. http://www.washingtontimes.com/national/20030609-122709-8176r.htm.

Jehl, Douglas. "Insiders' New Firm Consults on Iraq." *New York Times,* September 30, 2003. http://www.nytimes.com/2003/09/30/politics/30LOBB.html.

Johnson, Steve. "Critics Point to Snoop Factor in Airline Security." *Detroit Free Press,* June 10, 2003. http://www.freep.com/money/business/capps10_20030610.htm.

————. "Suit Seeks Details of Airline Passenger Screening Network." *San Jose Mercury News,* June 13, 2003. http://www.siliconvalley.com/mld/silicon valley/6079142.htm.

Journo, Elan. "Under Attack—by Eco-Terrorists: Homeland Security, Stopping the Destruction of Property in Defense of Nature." *San Francisco Chronicle,* October 12, 2003, sec. D. http://www.sfgate.com/cgi-bin/ article.cgi?file=/chronicle/archive/2003/10/12/INGET28F3B1.DTL.

Jung, Carl. *Memories, Dreams, Reflections.* Recorded and edited by Aniela Jaffé, translated by Richard and Clara Winston. New York: Vintage Books, 1963.

Kady, Martin. "Matrics Bets Chips on $14M Launch." *Washington Business Journal,* January 14, 2002. http://washington.bizjournals.com/washington/ stories/2002/01/14/story3.html.

Kaluszynski, Martine. "Republican Identity." In *Documenting Individual Identity: The Development of State Practices in the Modern World,* edited by Jane Caplan and John Torpey. Princeton, N.J.: Princeton University Press, 2001.

Kanellos, Michael. "Nanotech Spending Nears $3 Billion." *CNET News,* June 25, 2003. http://news.com.com/2100-1008_3-1021129.html.

Kerr, Jennifer. "Group Gets Private Data on U.S. Officials." Associated Press, August 28, 2003. http://story.news.yahoo.com/news?tmpl=story&u=/ap/ 20030828/ap_on_hi_te/privacy/concerns_5.

Kinder Investigations & K-9 Drug Detection Services. "Amphetamines." http://www.k9investigations.com/amphetamines.htm (accessed September 2, 2003).

Klare, Michael T. *Resource Wars: The New Landscape of Global Conflict.* New York: Metropolitan Books, 2001.

Klein, Bruce J. "Building a Bridge to the Brain: Researchers Are Close to Breakthroughs in Neural Interfaces, Meaning We Could Soon Mesh our Minds with Machines." Betterhumans, March 3, 2003. http://www.better humans.com/Features/Reports/report.aspx?articleID=2003-03-02-3.

Kover, Amy. "A Tech Company Wins Big in the War on Terror." *New York Times,* September 14, 2003. http://www.nytimes.com/2003/09/14/ technology/14COMM.html.

Krane, Jim. "Pentagon Could Debut New Weapons in Iraq." Associated Press, February 18, 2003.

Krebs, Brian. "Ex-Officials Urge U.S. To Boost Cybersecurity." *Washington Post,* April 9, 2003, sec. E.

Kren, George M., and Leon Rappoport. *The Holocaust and the Crisis of Human Behavior.* New York: Holmes & Meier, 1980.

Kurzweil, Ray. "We Are Becoming Cyborgs." *KurzweilAI.net.* http://www.kurzweilai.net/meme/frame.html?main=/articles/art0449.html (accessed September 8, 2003).

L-3 Communications. "L-3 Communications to Participate in $498 Million Contract to Provide Air Force Base Security Systems." *Business Wire,* September 4, 2003. http://search.hoovers.com/free/co/news/detail.xhtml ?COID=89493&ArticleID=NR20030904290.2_c23d0016273183b6.

La Boétie, Éttiene de. *Discours de la Servitude Volontaire.* Written in 1552. Translated by Harry Kurz as *The Politics of Obedience: The Discourse of Voluntary Servitude* (Buffalo, N.Y.: Black Rose Books, 1997). Our quote comes from the abridged and edited version at Personal Empowerment Resources, http://www.mind-trek.com/ (accessed November 1, 2003).

Laing, R. D. *The Politics of Experience.* New York: Ballantine Books, 1977.

Landay, Jonathan S. "Bold U.S. Strike in Yemen Kills 6." *Mercury News* (San Jose, Calif.), November 5, 2002. http://www.bayarea.com/mld/mecurynews/news/4446720.htm.

Landesman, Peter. "Arms and the Man." *New York Times,* August 17, 2003. http://www.nytimes.com/2003/08/17/magazine/17BOUT.html.

Lao-Tzu, *Taoteching.* Translated by Red Pine. San Francisco: Mercury House, 1996.

Lee, Jennifer. "Passports and Visas to Add High-Tech Identity Features." *New York Times,* August 24, 2003. http://www.nytimes.com/2003/08/24/national/24IDEN.html.

Lenk, Klaus. "Information Technology and Society." In *Microelectronics and Society: For Better or Worse,* edited by Gunter Friedrichs and A. Schaff. Oxford, UK: Pergamon Press, 1982.

Locke, John. "Second Treatise, Sections 138–40." In *Two Treatises of Government,* edited by Peter Laslett. New York: Mentor Books, New American Library, 1965.

Loeb, Vernon. "Concrete-Piercing Bombs Hammer Caves." *Washington Post,* December 13, 2001. http://www.globalsecurity.org/org/news/2001/011213-attack02.htm.

Lucassen, Leo. "A Many-Headed Monster." In *Documenting Individual Identity: The Development of State Practices in the Modern World,* edited by Jane Caplan and John Torpey. Princeton, N.J.: Princeton University Press, 2001.

Lydersen, Kari. "Spying for Fun and Profit." *AlterNet,* May 28, 2003. http://www.alternet.org/story.html?StoryID=16009.

Machado, Antonio. *Times Alone: Selected Poems of Antonio Machado.* Translated and selected by Robert Bly. Middleton, Conn: Wesleyan University Press, 1983.

MacNeil/Lehrer Online NewsHour. "The USA PATRIOT Act." http://www.pbs.org/newshour/bb/terrorism/homeland/patriotact.html (accessed June 30, 2003).

Maguire, James. "Who's Spying on You at Work?" *NewsFactor.com,* July 18, 2003. http://story.news.yahoo.com/news?tmpl=story2&u=/nf/20030718/tc_nf/21925&e=5.

Martin, Al. "Coming Soon: Flying Fascism on Your Doorstep." *Pravda,* February 20, 2002. http://english.pravda.ru/columnists/2002/02/20/26531.html.

Marx, Karl, and Friedrich Engels. *Manifesto of the Communist Party.* 1848. http://www.marxists.org/archive/marx/works/1848/communist-manifesto/.

Matsumoto, Craig. "Intel Envisions Smart Dust." *Light Reading,* March 20, 2003. http://www.lightreading.com/document.asp?doc_id=30018.

Mayer, J.P. *Max Weber and German Politics,* 2nd ed. London: Faber and Faber, 1956.

McCue, Andy. "Gillette Shrugs Off RFID-tracking Fears." *New York Times,* August 15, 2003. http://www.nytimes.com/cnet/CNET_2100-1039_3-5063990.html?ex=1061959699&ei=1&en=e3f11c9e48afdafd.

McCullagh, Declan. "Are Spy Chips Set to Go Commercial?" *ZDNet,* January 13, 2003. http://zdnet.com.com/2100-1107-980345.html.

———. "U.S. Could Deny GPS to Taliban." *Wired News,* October 20, 2001. http://www.wired.com/news/conflict/0,2100,47739,00.html.

McDonough, Brian. "Networked Computer Sensors Infiltrate Everything." *NewsFactor.com,* June 5, 2002. http://sci.newsfactor.com/perl/story/18088.html.

McGoey, Chris E. "Shoplifting: Racial Profiling." http://www.crimedoctor.com/shoplifting5.htm (accessed August 17, 2003).

McNeill, William H. *The Pursuit of Power: Technology, Armed Force, and Society since A.D. 1000.* Oxford, UK: Basil Blackwell, 1983.

McRandle, James H. *The Track of the Wolf: Essays on National Socialism and its Leader, Adolf Hitler.* Evanston, Ill.: Northwestern University Press, 1965.

Meek, James Gordon. "Ashcroft Tour to Plug Terror Bill." *New York Daily News,* August 6, 2003. http://www.nydailynews.com/08-06-2003/news/wn_report/story/106872p-96686c.html.

Menn, Joseph. "Techies, Politics Now Click." *Los Angeles Times,* August 11, 2003. http://www.latimes.com/business/la-fi-geeks11aug11000416,1,368 7057.story?coll=la-headlines-business-manual.

Merchant, Carolyn. *The Death of Nature: Women, Ecology and the Scientific Revolution*. San Francisco: HarperSanFrancisco, 1983.

Merton, Thomas. "Devout Meditation on Adolph Eichmann." In *Raids on the Unspeakable*. New York: New Directions, 1966.

Michels, Robert. *Political Parties: A Sociological Study of the Oligarchical Tendencies of Modern Democracy*. Translated by Eden Paul and Cedar Paul. New York: The Free Press, 1915.

Miller, Nathan. *Stealing from America: A History of Corruption from Jamestown to Reagan*. New York: Paragon House, 1992.

Mills, C. Wright. *The Causes of World War Three*. London: Secker & Warburg, 1958.

Mokhiber, Russell. *Corporate Crime and Violence*. San Francisco: Sierra Club Books, 1988.

Mother Jones. "Prison Spending Growing Six Times Faster than Higher Education Spending." July 11, 2001. http://www.motherjones.com/about_us/pressroom/prisons_release.html.

Mumford, Lewis. *The City in History: Its Origins, Its Transformations, and Its Prospects*. New York: Harbinger, 1961.

———. "Drama of the Machines." *Scribner's Magazine,* August 1930. In *Interpretations and Forecasts: 1922-1972*. New York: Harcourt Brace Jovanovich, 1973.

———. "Marx: Dialectic of Revolution." In *Interpretations and Forecasts: 1922–1972*. New York: Harcourt Brace Jovanovich, 1973.

———. *The Myth of the Machine, Volume I: Technics and Human Development*. New York: Harvest/HBJ, 1966.

———. "The Origins of War." In *Interpretations and Forecasts: 1922–1972*. New York: Harcourt Brace Jovanovich, 1973.

———. *The Pentagon of Power*. New York: Harcourt Brace Jovanovich, 1970.

National Emergency Number Association. "About/Contact NENA." http://www.nena9-1-1.org/About_Contact/index.htm (accessed September 14, 2003).

———. "NENA Business Alliance Committee Members Announced." News release, September 11, 2000. http://www.nena9-1-1.org/PR_Pubs/Press Releases/Member%20News%20Release—NBA%209-11-00.PDF.

Newberg, Andrew, Eugene D'Aquili, and Vince Rause. *Why God Won't Go Away: Brain Science and the Biology of Belief*. New York: Ballantine Books, 2002.

NOVA. *Spies That Fly*. Public Broadcasting System (PBS). http://www.pbs.org/wgbh/nova/spiesfly/.

O'Harrow Jr., Robert. "Air Security Network Advances." *Washington Post,* March 1, 2003, sec. E. http://www.washingtonpost.com/ac2/wp-dyn?page name=article&node=&contentId=A18601-2003Feb28&notFound=true.

———. "U.S. Backs Florida's New Counterterrorism Database." *Washington Post,* August 6, 2003, sec. A.

Orwell, George. *1984.* New York: Signet Classics, 1962.

Paez, Angel. "CIA Gave $10 Million to Peru's Ex-Spymaster." *The Public i,* July 3, 2001.

Pahati, Omar J. "Confounding Carnivore: How to Protect Your Online Privacy." *AlterNet,* November 29, 2001.

Parenti, Christian. *Lockdown America: Police and Prisons in the Age of Crisis.* London and New York: Verso Books, 1999. http://www.thirdworldtraveler .com/Prison_System/BigBucks_BigHouse_LA.html (accessed September 22, 2003).

Pernoud, Regine. *Those Terrible Middle Ages!* Translated from the 1977 French edition by Anne Englund Nash. San Francisco: Ignatius, 2000.

Pert, Candace B. *Molecules of Emotion: The Science Behind Mind-Body Medicine.* New York: Touchstone, 1997.

Piller, Charles. "Army of Extreme Thinkers." *Los Angeles Times,* August 14, 2003. http://www.latimes.com/news/science/la-sci-darpa14aug14,1,6143132 .story?coll=la-news-science.

Pincus, Craig. "U.S. Strike Kills Six in Al Qaeda." *Washington Post,* November 5, 2002. http://www.washingtonpost.com/ac2/wp-dyn/A5126-2002Nov4.

Platt, Charles. "The Museum of Nanotechnology." *Wired.* http://www.wired .com/wired/scenarios/museum.html (accessed August 17, 2003).

Pollack, Andrew. "Army Center to Study New Uses of Biotechnology." *New York Times,* August 27, 2003. http://www.nytimes.com/2003/08/27/national/ 27BIOT.html.

Prendergast, William. Nanotechnology patent lawyer and partner in the Chicago law firm of Brinks, Hofer, Gilson & Lione. August 28, 2003. http://www.etcgroup.org/main.asp (accessed October 22, 2003).

Press Association. "DNA Database Stores 2m Profiles." *The Guardian* (UK), June 25, 2003. http://www.guardian.co.uk/crime/article/0,2763,984753,00 .html (accessed August 14, 2003).

Priest, Dana. "U.S. Citizen Among Those Killed In Yemen Predator Missile Strike." *Washington Post,* November 8, 2002.

Rachel's Environment and Health News. "The Revolution. Part 1." June 26, 2003. http://www.rachel.org/bulletin/bulletin.cfm?Issue_ID=2362.

————. "The Revolution. Part 2." July 10, 2003. http://www.rachel.org/
bulletin/bulletin.cfm?Issue_ID=2363.

————. "The Revolution. Part 3." July 24, 2003. http://www.rachel.org/bul-
letin/bulletin.cfm?Issue_ID=2371.

Radford, Tim. "Brave New World or Miniature Menace?" *The Guardian*
(UK), April 29, 2003. http://www.guardian.co.uk/uk_news/story/0,3604
,945498,00.html.

Reddy, Anitha. "U.S. Readies Program to Track Visas." *Washington Post,*
September 29, 2003. http://www.washingtonpost.com/wp-dyn/articles/
A14287-2003Sep28.html.

Reporters Committee for Freedom of the Press. "Homefront Confidential:
How the War on Terrorism Affects Access to Information and the Public's
Right to Know." 4th edition, September 2003. http://www.rcfp.org/home
frontconfidential/foi.html.

Reuters. "Robot Shows Prime Minister How to Loosen Up." August 22, 2003.

Rosen, Cheryl, and Mathew G. Nelson. "The Fast Track." *InformationWeek,*
June 18, 2001. http://www.informationweek.com/shared/printableArticle
?doc_id=IWK20010618S0001.

Ross, Colin A. *Bluebird: Deliberate Creation of Multiple Personality by
Psychiatrists*. Richardson, Tex.: Manitou Communications, 2000.

Rousseau, Jean-Jacques. *On the Social Contract*. Mineola, N.Y.: Dover
Publications, 2003.

Rybcynski, Witold. *Taming the Tiger: The Struggle to Control Technology*. New
York: Viking Press, 1983.

Scheeres, Julia. "Kidnapped? GPS to the Rescue." *Wired News,* January 25,
2002. http://www.wired.com/news/business/0,1367,50004,00.html.

————. "The R's: Reading, Writing, RFID." *Wired News,* October 24, 2003.
http://www.wired.com/news/technology/0,1282,60898,00.html?tw=wn_tech
head_5.

Schmidt, Charles W. "The Networked Physical World." http://www.rand.org/
scitech/stpi/ourfuture/Internet/sec4_networked.html.

Schneier, Bruce. "Terror Profiles By Computers Are Ineffective." *Newsday,*
October 21, 2003. http://www.newsday.com/news/opinion/ny-vpsch
213503428oct21,0,3927478.story.

Scott, James. *Domination and the Arts of Resistance: Hidden Transcripts*. New
Haven, Conn.: Yale University Press, 1990.

————. *Weapons of the Weak: Everyday Forms of Peasant Resistance*. New
Haven, Conn.: Yale University Press, 1985.

Seal, Cheryl. "Frankensteins in the Pentagon: DARPA's Creepy

Bioengineering Program." *Information Clearinghouse: News You Won't Find on CNN*. http://www.informationclearinghouse.info/article4572.htm (accessed September 1, 2003).

Seisint. "Seisint Overview." http://www2.seisint.com/aboutus/index.html (accessed September 24, 2003).

Shachtman, Noah. "Rage Against the (Green) Machine." *Wired News*, June 19, 2003. http://www.wired.com/news/technology/0,1282,59287,00.html.

Shen, Michelle. "The 'People' Element In Biometrics And Physical Access Control." *BioMetriTech*, April 14, 2003. http://www.biometritech.com/features/shen041403.htm.

———. "Trends In Biometric Security (Part 3): Buyer Behavior Analysis." *BioMetriTech*, March 19, 2003. http://www.biometritech.com/features/shen031903.htm.

Shenon, Philip. "Former Domestic Security Aides Switch to Lobbying." *New York Times*, April 29, 2003. http://www.nytimes.com/2003/04/29/politics/29HOME.html.

Smith, Adam. *An Inquiry into the Nature and Wealth of Nations: A Selected Edition*. Edited by Kathryn Sutherland. Oxford: Oxford University Press, 1993.

Smith, Derek V., Chairman and Chief Executive Officer of ChoicePoint. "Remarks at ChoicePoint 2003 Annual Shareholders Meeting," April 29, 2003. http://www.choicepoint.net/news/feature042903.html.

Smith, Marc. "Church History: Kill Them All, Let God Sort Them Out!" *Straitway*. http://straitway.org/2001/03012001.htm (accessed September 15, 2003).

Sniffen, Michael J. "Pentagon Developing High-tech Surveillance Project: Cameras, Computers Would Track Vehicles in Foreign Cities." *San Francisco Chronicle*, July 2, 2003, sec. A.

———. "Pentagon Developing System that Can Track Every Vehicle in a City." *CNEWS*, July 1, 2003. http://www.canoe.com/CNEWS/TechNews/2003/07/01/124436-ap.html.

———. "Proposed System Would Use Lots of Data." *The Guardian*, May 19, 2003.

Stanley, Jay, and Barry Steinhardt. *Bigger Monster, Weaker Chains: The Growth of an American Surveillance Society*. American Civil Liberties Union, 2003. http://www.aclu.org/Files/getFile.cfm?id=11572.

Sullivan, Andy. "Cutting-Edge Science Creates Stain-Free Pants." *USA Today*, July 23, 2003. http://www.usatoday.com/tech/news/techinnovations/2003-07-23-robopants_x.htm.

————. "Identity Theft Strikes 1 in 8 Adults, FTC Says." Reuters, *Yahoo News,* Asia, September 4, 2003. http://asia.news.yahoo.com/030903/3/13vwt.html (accessed September 5, 2003).

————. "Military Says Computer Dragnet to Include Limits." Reuters, May 20, 2003.

Sullivan, Bob. "Why We're All at Risk of ID Theft." *MSNBC,* January 21, 2003. http://www.msnbc.com/news/758896.asp (accessed September 5, 2003).

Sun-Sentinel (Fort Lauderdale, Fla.). "State Law Enforcement Contractor Linked To Drugs." August 3, 2003. http://www.mapinc.org/drugnews/v03/n1171/a06.html.

Sunshine Project. "Ethnically-specific Biological Weapons: An Analysis of Human Genome Data Reveals That Ethnically-specific Genetic Markers Do Exist." Briefing paper from *Emerging Technologies: Genetic Engineering and Biological Weapons.* Sunshine Project Backgrounder No. 12, October 2003. http://www.sunshine-project.org/publications/others/snpbw.html (accessed October 25, 2003).

Suppes, Patrick, Bing Han, and Zhong-Lin Lu. "Brain-wave Recognition of Sentences." *Proceedings of the National Academy of Sciences* 95, no. 26 (December 22, 1998): 15861–66. http://www.pnas.org/cgi/content/full/95/26/15861.

Sybase. "About Sybase." http://www.sybase.com/about_sybase (accessed August 19, 2003).

Taylor, Frederick Winslow. *Principles of Scientific Management, Comprising Shop Management, The Principles of Scientific Management and Testimony Before the Special House Committee.* New York: Harper & Row, 1911.

Traven, B. *The Death Ship.* Brooklyn, N.Y.: Lawrence Hill, 1991. Originally published as *Totenschiff* (Berlin: Buchmeister Verlag, 1926).

UNICOR. *2002 Annual Report, Statements of Operations and Cumulative Results of Operations.* http://www.unicor.gov/about/2002annual/auditors_report05.htm (accessed September 22, 2003).

University of Southern California News Service. "Machine Demonstrates Superhuman Speech Recognition Abilities." News release 0999025, September 30, 1999. http://www.fas.org/irp/program/process/36013.htm (accessed September 13, 2003).

U.S. Air Force Scientific Advisory Board. *New World Vistas: Air and Space Power for the 21st Century.* Washington, D.C.: USAF Scientific Advisory Board, 1995-96. 15 volumes. The summary volume can be found online at http://www.fas.org/spp/military/docops/usaf/vistas/vistas.htm. The ancillary

volume we quoted can be found at http://stinet.dtic.mil/cgi-bin/fulcrum
_main.pl?database=ft_u2&searchid=10677104835631&keyfieldvalue=ADA
309597&filename=%2Ffulcrum%2Fdata%2FTR_fulltext%2Fdoc%2FADA
309597.pdf.

U.S. Central Intelligence Agency. *KUBARK Counterintelligence Interrogation Manual.* July 1963.

U.S. Congress. House. Hearing before the Committee on Appropriations, Subcommittee on the Departments of Commerce, Justice and State, the Judiciary and Related Agencies. February 28, 2002 (testimony of U.S. Attorney General John Ashcroft). http://www.usdoj.gov/ag/testimony/ 2002/FY2003AG_WrittenStatement-House.htm.

———. Joint House/Senate. Counterterrorism Information Sharing with Other Federal Agencies and with State and Local Governments and the Private Sector: Hearing before Joint House/Senate Intelligence Committee. October 1, 2002 (statement of Eleanor Hill). http://www.fas.org/irp/ congress/2002_hr/100102hill.html.

———. Senate. Select Committee to Study Governmental Operations with Respect to Intelligence Activities, *Final Report.* 94th Cong., 2d sess., 1976. S. Rep. 94-755. The text of Volume 2, *Intelligence Activities and the Rights of Americans,* which includes the committee's recommendations, can be found at http://www.icdc.com/~paulwolf/cointelpro/churchfinalreportIId.htm.

U.S. Defense Advanced Research Projects Agency (DARPA). "Ability to Simulate Gunshot Wounds Provides Realistic Training." March 27, 2001. http://www.darpa.mil/body/legacy/prev_items.html.

———. "Autonomous Vehicles Grand Challenge." http://www.darpa.mil/ grandchallenge/overview.htm (accessed August 11, 2003).

———. "BioSPICE Project Proposal Solicitation." http://www.darpa.mil/ito/ Solicitations/PIP_01-26.html. (accessed July 1, 2003).

———. "Harvesting Biology for Defense Technology Conference, June 23–25, 2003." Card at http://web-ext2.darpa.mil/body/ppt/DARPAPost CARD.PPT (accessed September 1, 2003).

———. "Legacy." http://www.darpa.mil/body/legacy/prev_items.html (accessed August 18, 2003).

U.S. Department of Defense. "Assistant Secretary of Defense for Networks and Information Integration/Department of Defense Chief Information Officer." http://www.dod.mil/nii/bio/asd/ (accessed September 13, 2003).

U.S. Department of Defense, Biometrics Management Office. "Frequently Asked Questions." https://www.bfc-kno.army.mil/faq/faq.htm (accessed August 7, 2003).

U.S. Department of Justice. Bureau of Justice Statistics. "Expenditure and Employment Statistics." http://www.ojp.usdoj.gov/bjs/eande.htm (accessed September 22, 2003).

U.S. Department of Transportation, Office of the Secretary. *Privacy Act of 1974: System of Records.* "Notice to amend a system of records." In *Federal Register* 68, no. 10 (January 15, 2003): 2101–2103.

U.S. Federal Bureau of Investigation. "Carnivore: Diagnostic Tool." http://www.fbi.gov/hq/lab/carnivore/carnivore2.htm (accessed October 28, 2003).

U.S. Federal Trade Commission. *Identity Theft Report.* Prepared for FTC by Synovate. McLean, Va.: Synovate, September 2003. http://www.ftc.gov/os/2003/09/synovatereport.pdf.

U.S. National Institute of Standards and Technology. "Biometric Consortium Conference, September 23–25, 2002. About Biometrics." http://www.itl.nist.gov/div895/isis/bc/bc2002/aboutbiometrics.htm (accessed September 13, 2003).

———. "Biometric Interoperability, Performance, and Assurance Working Group." http://www.itl.nist.gov/div895/isis/bc/bcwg/ (accessed September 13, 2003).

———. "General Information." http://www.nist.gov/public_affairs/general2.htm (accessed September 13, 2003).

———. "NIST in Your House." http://www.nist.gov/public_affairs/nhouse/index.html (accessed September 13, 2003).

———. "Technologies for Improved Homeland Security." http://www.nist.gov/public_affairs/factsheet/homeland.htm (accessed September 8, 2003).

Valigra, Lori. "Fabricating the Future." *Christian Science Monitor,* August 29, 2002. http://csmweb2.emcweb.com/2002/0829/p11s01-stgn.html.

VanScoy, Kayte. "Can the Internet Hot-Wire P&G?: They Know What You Eat." *Ziff Davis Smart Business,* January 1, 2001.

VeriChip. "Products and Services. VeriChip." http://www.adsx.com/prodservpart/verichip.html (accessed September 13, 2003).

Viladas, Pilar. "Home Despot." *New York Times Magazine,* August 17, 2003. http://www.nytimes.com/2003/08/17/magazine/magazinespecial/WFOTHOMET.html.

Wakefield, Jane. "US Looks to Create Robo-soldier." *BBC News,* April 10, 2002. http://news.bbc.co.uk/1/hi/sci/tech/1908729.stm (accessed June 21, 2003).

Warneke, Brett. "Smart Dust." University of California, Berkeley, Department of Electrical Engineering and Computer Sciences. http://www-bsac.eecs.berkeley.edu/~warneke/SmartDust/index.html (accessed August 11, 2003).

Washington, George. First Inaugural Address (April 30, 1789). Published at *The Avalon Project at Yale Law School.* http://www.yale.edu/lawweb/ avalon/presiden/inaug/wash1.htm (accessed June 14, 2003).

Washington Post editorial. "What Is Operation TIPS?" July 14, 2002. http://www.washingtonpost.com/ac2/wp-dyn?pagename=article&node =&contentId=A63924-2002Jul12&notFound=true.

Watson, David. *Against the Megamachine: Essays on Empire and its Enemies.* New York: Autonomedia, 1998.

Weber, Max. *Economy and Society.* Edited by Guenther Roth and Claus Wittich. Translated by Ephraim Fischoff and others. New York: Bedminster Press, 1921, 1968.

———. *From Max Weber.* Translated and edited by H. H. Gerth and C. Wright Mills. New York: Galaxy, 1946, 1958.

———. *The Protestant Ethic and the Spirit of Capitalism.* Translated by Talcott Parson. New York: Charles Scribner's Sons, 1904, 1930.

Weiser, Benjamin. "F.B.I. Accused of Corrupting Computer Surveillance." *New York Times,* August 20, 2003. http://www.nytimes.com/2003/08/20/ nyregion/20STEW.html.

Welsh, Cheryl. "Best Mind Control Documentary Excerpts." http://www.dcn.davis.ca.us/~welsh/tvlist.htm (accessed May 20, 2003).

World Summit on the Information Society. "Basic Information: About WSIS." http://www.itu.int/wsis/basic/about.html (accessed November 9, 2003).

———. "Information on Business Input." http://www.iccwbo.org/home/ e_business/wsis.asp (accessed November 9, 2003).

World Tribune. "In Wake of Predator Success, U.S. Weighs Assassination Options." November 29, 2002. http://216.26.163.62/2002/ss_terrorism _11_29.html.

Yannuzzi, Rick E. "In-Q-Tel: A New Partnership Between the CIA and the Private Sector." *Defense Intelligence Journal* 9, no. 1 (Winter 2000). http://www.cia.gov/cia/publications/inqtel/.

Index

A

Abbey, Edward, 39
accountability, 138
ACLU (American Civil Liberties Union), 174–75, 213
advertising, 79
Afghanistan invasion, 2001, 179–80
aggression, 78, 80
airport security, 128–31, 163, 183
Albrecht, Katherine, 5
Alien Technology Corporation, 153
all-seeing eye, 1–13, 18–19, 176. *See also* Panopticon; RFID (Radio Frequency Identification tags)
Alvares, Claude, 14
American Civil Liberties Union (ACLU), 174–75, 213
American Indians, 12, 98–99, 219–20
Amsec, 165
anthropometry, 115–16, 139
Arendt, Hannah, 203
arms merchants, 162–63
Aronowitz, Stanley, 24, 30–32, 40–41
Ashcroft, John, 126–27, 158, 159, 176
Asher, Hank, 161
Auto-ID. *See* RFID (Radio Frequency Identification) tags
automation, 109–10
Axciom Corporation, 129, 151, 175

B

Barnett, Martha, 162
Bauman, Zygmunt, 221
Bellah, Robert N., 83
Benneton, 154
Bentham, Jeremy, 8–10, 92, 115, 119, 140
betterhumans.com Web site, 60–63
Biloxi, Mississippi, student monitoring program, 36
biological weapons, 17, 41–42, 46–47
Biometric Consortium, 183
Biometric Interoperability, Performance and Assurance Working (BIPAW) Group, 183–84
biometrics, 139, 148, 164, 183–86

brain fingerprinting, 185–86
definition, 117–18
Biometrics Market Intelligence, 150–51
biotechnology, 16–17, 45–49, 182–83, 189
Bolonkin, Alexander, 63–69, 72, 91
Bout, Victor, 162–63
brain fingerprinting, 185–86
Brain Interface Program, 15, 47
Britain
    ECHELON computer surveillance, 169
    national DNA database, 120–21
bureaucracies, 28, 118, 119, 132, 215
    accountability in, 138
    communities, effect on, 73–74, 102
    compartmentalization of responsibilities, 123
    corporate, 28, 125, 131
    definition and overview, 73, 79, 101, 125–26
    discipline, use of, 133–34, 139
    efficiency of, 79, 101–104
    hierarchy in, 74, 101
    knowledge, use of, 133
    mechanical behavior in, 106–107
    power and control in, 73–74, 100–103, 115, 133, 213
    sorting of society by, 197
Bush, George W., and family, 165–66, 176–77

C

California Anti-Terrorism Information Center, 149
cancers, 88–89
capitalism, 85, 94, 96, 142
Caplan, Jane, 147
CAPPS (Computer Assisted Passenger Prescreening System), 128–30
Carlyle Group, The, 153, 166
Carnivore packet analyzer, 171–74
cars, tracing location of, 121–24
Catholic Church, 85–86
cell phones, tracing location of, 121–22
Center for Embedded Network Sensing (CENS), 177–78

Central Intelligence Agency. *See* U. S.
    Central Intelligence Agency (CIA)
chemical weapons, 17
children
    abuse of, 90–91
    fingerprinting of, 7–8, 183
    high school, police raid at, 37–38
    RFID implants in, 156
    student identification tags, 36–37
    student monitoring program, 36
ChoicePoint, 151, 164, 175
Christianity. *See* religion
Chuang tzu, 142
Churchill, Ward, 54–55
CIA. *See* U. S. Central Intelligence Agency
    (CIA)
citizen information, 23, 112, 176. *See also*
    identity cards; individuals, information
    on; passports
citizen participation in law enforcement,
    159, 201–203
civilized and indigenous ways of being,
    compared, 82–85
communications, surveillance of, 169–74.
    See also e-mail, monitoring of
Communist Manifesto, The, 92–96
communities, 43, 181, 216–17
    definition, 145
    destruction of, 142, 146
    privacy in, 208–209
Computer Assisted Passenger Prescreening
    System (CAPPS), 128–30
computers, 63, 72–73, 217. *See also* nano-
    technology
    e-mail, monitoring of, 149, 169–74
    military use, 182
    passwords, FBI collection of, 173
conformity. *See* individuals as cogs in
    machine
consent, governance by, 195–97
consumer information, 22–23, 112, 147,
    149. *See also* individuals, information
    on; RFID (Radio Frequency
    Identification) tags
consumerism, 94, 100, 141, 142
control, 28, 33–49, 109, 141, 168, 217.
    See also power
    of bureaucracies (*See* bureaucracies)
    paranoia, effect of, 78

productivity and, 79
science and control of natural world, 40,
    41, 73
surveillance and, 117, 126, 132
technology and, 112, 210
corporations
    accountability in, 138
    affiliates, sharing of information by, 53
    as bureaucracies (*See* bureaucracies)
    crimes of, 166–67
    government privatization and, 151, 175
    government-sponsored technologies, role
    in, 71–72
    maximizing profits for, 91
    power of, 198
Cummings, Philip, 51
customer surveillance, 154–56
cyberspace, securing, 148, 173–74

D

DARPA. *See* Defense Advanced Research
    Projects Agency (DARPA)
data creep
    air passenger screening, 129
    car event data recorders, 123
DBT Online company, 162
DCHDs (Domestic Control Hover
    Drones), 180–81
death and immortality, 60–64, 80–81
Defense Advanced Research Projects
    Agency (DARPA), 18–19, 45–48, 180,
    188–89, 194, 209
deforestation, 106
Deloria, Vine, 82–85
democracy, 28, 81, 101, 141, 198
discipline, use by bureaucracy of, 133–34,
    139
diversity, destruction of, 75–76, 79
DNA databases, 164
    British national, 120–21
    U. S. Combined DNA Index System, 121
Domestic Control Hover Drones
    (DCHDs), 180–81
Drexler, Eric, 70–71
Durkheim, Emily, 132

E

ECHELON computer software, 169–70
ecological understanding, 95

economic incentives to provide
information, 141
ecosystems, 144–45
E-creatures, 63–69
efficiency, 28, 81, 107
of bureaucracies, 79, 101–104
of machines, 74–75
of Panopticon guards, 157
e-mail, monitoring of, 149, 169–74
employment information, availability of,
22, 112, 147, 176
Engels, Friedrich, 92–96
Equifax, 164
ethnic identity cards, 114
ethnically-specific biological weapons,
41–42
Euro banknotes, RFID tags in, 6
event data recorders, car, 122–24
exoskeletons, for soldiers, 14
Experian, 164

F
face recognition technology, 36, 117, 149
false choices, 29
FBI. *See* Federal Bureau of Investigation
(FBI)
fear, 78–87, 187
advertising, as basis of, 79
commercial response to, 163
of death, 80–81
security, obsession with, 127, 142
technologies to inhibit, 48
Federal Bureau of Investigation (FBI)
Carnivore packet analyzer, use of,
171–74
National Instant Check System, 157–58
Sesint and, 161–62
surveillance by, 149, 175–76, 201–202,
215
financial information, 22, 127.
See also identity theft; individuals,
information on
fingerprinting
brain, 185–86
children, 7–8, 183
passports and identity cards, 115–16, 118,
148
welfare recipients, 119–20
forests, destruction of, 106

Foucault, Michel, 9–10, 92, 132, 133
France, Anatole, 134
freedom, 28, 126, 193, 197, 208
Freedom of Information Act requests,
176–77, 179
Fromm, Erich, 88
Fukuyama, Francis, 31

G
gait identification, 17
Gandy, Oscar, 112
General Motors, 123
genetic engineering, 41–42, 46, 59
genius, 143
genocide, 99, 106–107, 116–17, 193
George, W. L., 114–15
Georgia Institute of Technology, 17
Gillette Corporation, 6, 153
global ecophagy, 70–71
global information society, 198–200
global positioning system (GPS)
technology, 122, 124
global warming, 106
globalization, 94
Goodyear Tire and Rubber Company,
151–52
government benefit recipients, information
available on, 23
government bureaucracies. *See* bureaucracies
government information, secrecy of,
176–77
government/corporate interconnection, 198
See also bureaucracies; corporations
government-subsidized technologies,
five-step pattern in, 71–72
Graham, Bob, 161
greed, 78, 141, 142, 163
Greneker, Gene, 17
Gruen, Arno, 33, 50, 187

H
Heyman, Charles, 17
hierarchical relationships, 74, 77, 124, 172,
174
Hilberg, Raul, 106
Hirst, Paul, 16
Hitachi Europe, 6
Hollerith computers, 116–17

Holocaust, the, 106–107, 116–17
*Human Resource Exploitation Training
    Manual,* 99

I

ICC (International Chamber of
    Commerce), 200–201
identity, 50–56. *See also* individuals, infor-
    mation on; privacy
  in communities, 208
  ownership of, 206, 207
identity cards and passports, 113–18, 148,
    150
identity theft, 51–56, 143
  corporate affiliates, information shared
    by, 53
  Cummings case, 51
  financial information, 51–53, 143
  giving ourselves away, 54–55, 143
  Panopticon, effect of, 55–56, 81
  schooling, role of, 54
immigration identification, 117, 150
immortality and death, 60–64, 80–81
indigenous peoples, 87, 98–99, 142. *See also*
    American Indians
  civilized *vs.* indigenous ways, 82–85, 89,
    144
  science and, 84, 88
individuals, information on, 22–23, 112.
    See also biometrics; fingerprinting;
    identity theft; monitoring of individ-
    uals; privacy; RFID (Radio Frequency
    Identification) tags; Total Information
    Awareness (TIA)
  air passenger screening, 128–31
  DNA databases, 120–21
  economic incentives, 141, 147
  government control of, 215
  identity cards and passports, 113–18, 148,
    150
  power gained through, 213
  privatization, effect of, 151, 164–65, 175
  security incentives, 141
  welfare recipient identification,
    Connecticut, 119–20
individuals as cogs in machine
  acceptance of role, 136–40, 143–45,
    193–94, 197–98
  conformity, pressure for, 81, 97–98

maintenance of machine, individual's
    role in, 74–75, 212
  stopping the machine, 213–16
industrial espionage, 169
industrial state, 126
  bureaucracies, role of, 101–104
  democracy in, 101
  globalization of, 94
  landbases destroyed by, 28, 142, 161
  Marx and Engels' analysis of, 92–96
  Mumford's analysis of, 108–11
  panoptic sort in, 147
information, personal. *See* individuals,
    information on
Information Awareness Office, U.S.,
    18–19, 23–24
information liquidity, 165
Innocent III, Pope, 86
In-Q-Tel, 167
Institute for Collaborative Biotechnologies,
    182–83
intellectual property, 206–208
International Chamber of Commerce
    (ICC), 200–201
International Civil Aviation Organization,
    118
International Convention on Cybercrime,
    174
International Labour Organization, 118
Internet, 173–74, 182. *See also* e-mail,
    monitoring of
interrogation techniques, 99–100, 107–108,
    119, 121, 127–28, 131–32, 158
Iraq War, 37, 180
Iron Law of Oligarchy, 101

J

job satisfaction, 89
Johnson, Roy, 188
Jung, Carl, 78

K

Kaluszynki, Martine, 138–39
Kass, Leon, 48
knowledge
  bureaucracy, use of, 133
  power and, 19–25, 39, 168, 209–10
  science defined as, 23–25, 85
  as value-free, 31

KUBARK Counterintelligence Interrogation Manual, 99, 107–108, 119, 121, 127–28, 131–32

L

L-3 Communications, 163
La Boétie, Éttiene de, 195–97, 221
Laing, R. D., 8
landbases, 94, 161
    destruction of, 28, 142, 161
    livability of, 89
Lao-tzu, 215–16
laws and rules, 207–209. *See also* privacy laws
    relationships in place of, 209, 217
    in science, 40–41, 73, 84
    in society, 133–34, 139, 193
Lenk, Klaus, 212–13
Lockheed Martin Corporation, 128

M

Machado, Antonio, 1, 176
machine, the, 57–77, 96. *See also* bureaucracies; laws and rules; nanotechnology; Panopticon
    accountability in, 138
    behavior controlled by, 97–98
    community and, 144–45
    efficiency of, 74–75
    fear of death and, 80
    genius and, 143
    globalization of, 94
    individuals as cogs in (*See* individuals as cogs in machine)
    information elimination and, 91–92
    life converted into power, 213–14
    Marx and Engels' analysis, 93
    Mumford's analysis, 108–11
    neural nets, machine based on, 145–46
    social pyramid structure, 190–93
    social role of, 72–73
    stopping, 213–16
Macy's department stores, 148, 154
Magic Lantern virus, 173
magnetic resonance imaging (MRI) machines, 4
Marmon Group, 164–65
Marx, Karl, 92–96, 212

mass media, 126, 198–99
Massachusetts Institute of Technology (MIT), 6, 14, 148, 163, 182–83
Matrics company, 153
Matrix, The, 221–22
Matrix (Multistate Anti-Terrorism Exchange), 160
maturity as goal of life, 84–85
McGoey, Chris, 154–56
Meese, Edwin, 175
Merton, Thomas, 203
messiah, need for, 83
Michelin Tire, 151–52
middle class, 93–94, 113, 140
military use of science and technology, 178–84. *See also* U. S. Department of Defense
    biotechnology, 16–17, 45–49, 182–83
    Continuous Assisted Performance technology, 48
    fear, inhibition of, 48
    industrial development and, 94
    Persistence in Combat technology, 47–48
    übersoldiers, development of, 14–15
Mills, C. Wright, 107
MIT. *See* Massachusetts Institute of Technology (MIT)
molecular engineering. *See* nanotechnology
money, 142, 147–67. *See also* wealth
    RFID tags in, 6
    as root of all evil, 10–11, 146
    U.S. dollar, 11–12, 190
monitoring of individuals, 81, 87, 149, 169–74, 178
Montague, Peter, 71–72
multiple choice test, 44
Multistate Anti-Terrorism Exchange (Matrix), 160
Mumford, Lewis, 70, 108–11, 147, 190–93

N

nanobots, 63
nanoparticles, 59, 71
nanotechnology, 59–74
    global ecophagy, 70–71
    government, role of, 69, 71–72, 148, 149
    immortality as goal, 60–64
    Smart Dust, 177–78

nanotechnology (*continued*)
  übersoldiers, development of, 14
National Emergency Number Association
  (NENA), 122
National Highway Traffic Safety
  Administration (NHTSA), 123
National Instant Check System, 157–58
National Institute of Standards and
  Technology (NIST), 183–84
National Nanotechnology Initiative, 69
National Security Agency, 183, 215
nature, 28, 40, 41, 209–10
Nazis, 116–17, 203
Neighborhood Watch, 148
911 attack, 117, 128, 157–58, 163, 176,
  189–90
911 calls, identifying location of, 122

O

observation *vs.* objectivity, 39, 41
Oligarchy, Iron Law of, 101
opposites, world in terms of, 27–29
Orwell, George, 58, 125, 168, 209

P

panoptic sort, 112–24, 139, 140, 147, 197
  example of, 134–36
  judicial system, use of, 211
Panopticon, 12–13, 156, 185, 217, 219, 220,
  222, 224
  Bentham's design, 8–9, 140
  force, based on, 97
  Foucault's commentary, 9–10
  goods and services needed to operate,
    147–51
  government information, secrecy of,
    176–77
  hierarchical relationships in (*See* hierarchical relationships)
  identity, effect on, 55–56, 81
  military use, 17, 179
  nanotechnology and, 66, 67, 68, 72–73
  people as chattel in, 205–206
  power and, 133
  privacy and, 130, 168, 208
  pyramid structure of, 190–93
  rational behavior and, 89
  RFID, role of, 156–57

salvation and, 85, 87
shopping mall, panoptic architecture of,
  141
Smart Dust, use of, 177–78
stakeholders in, 199
technology, use of, 181–86, 195
paranoia, 78–79
passports, 113–17, 148
Patriot Act, 158, 165, 173
Patriot II Act, 127
Pelican Bay State Prison, 9, 26, 98
  SHU, 218–19
Pentagon. *See* U. S. Department of
  Defense
pesticide use, 88–89, 92, 182
photographs, identification, 115–16, 118
Pike, John, 36
Platform computer network, 169
Poindexter, John, 160
Pointer unmanned aerial vehicle (UAV),
  152–53
police, 28
  high school police action, 37–38
  normalizing of police violence, 38
pornography, 26–27, 173, 174
power, 86, 141. *See also* control
  bureaucracies and (*See* bureaucracies)
  changing power relations in society, 195
  concentration of, 94, 100, 140
  definition, 73
  false choices as device of, 29
  freedom undermined by, 208
  knowledge and information, role of,
    19–25, 39, 99–100, 112, 117, 168,
    209–10
  life converted into, 213–14
  of machines, 140, 143
  money as, 39, 146
  Panopticon and, 133
  technology and, 39, 182
Prada, 154
Predator drone, 178–80
President's Council on Bioethics, 48
prisons, 89, 98, 211, 218–19, 224. *See also*
  Pelican Bay State Prison
  costs of, 150
  federal prisons, expansion of, 159

privacy, 126, 223–24
  concerns for, 198
  power gained by information-gathering, 212–13
  property, relation to, 207–208
  security and, 29–30, 43–45, 174–76
  in traditional communities, 208–209
Privacy Act of 1974, 175–76
privacy laws, 118, 123, 126, 130, 174–76, 210
private property, 95
privatization, 151, 175
Procter & Gamble, 6, 152
productivity, 75, 79, 91
profiling, 154–56
Project for the New American Century, 42, 46
property, 207–208, 224
  public, fragmenting of, 43
pyramid social structure, 190–93

R
Rachel's Environment and Health News, 71
racial profiling, 154–56
radar devices, gait identification using, 17
Radio Frequency Identification tags. *See* RFID (Radio Frequency Identification) tags
Rand Corporation, 142, 184
rape, 58, 89–91
rational behavior, 96
  bureaucrats, rational discipline of, 133–34, 139
  of modern civilization, 88–89, 91, 203–204
rationalization, 88–111
rats, experiments on, 15, 47, 71
relationships, 90. *See also* hierarchical relationships
  science and, 28
  in traditional communities, 208–209
  unequal, 26–27, 90–91
religion, 40, 83–86, 94, 217, 219
  accumulation of wealth and, 96–97
  Catholic Church, 85–86
resources, human and nonhuman, 74–75, 94, 144
responsibility, 209

RFID (Radio Frequency Identification) tags, 5–6, 17, 34, 151–54, 156–57
  implants in humans, 156
  in student identification tags, 36–37
Ridge, Tom, 166
robots, 15–17, 194
  DARPA Los Angeles to Las Vegas robot race, 188, 194
  nanobots, 63
  Persistence in Combat technology, 47–48
  rats, control of, 15, 47
  unmanned aerial vehicles (UAVs), 152–53, 178–82
Roco, Mihail, 69–70
Rousseau, Jean-Jacques, 195
Rudolph, Alan S., 47
rules. *See* laws and rules
Rwanda genocide, 114
Rybcynski, Witold, 194–95

S
salvation, 83, 85, 87
schooling, effect on identity of, 54, 75
science, 14–32, 38–39, 94, 108, 132
  acceptance by population, 194
  brain fingerprinting, 185–86
  control and, 40, 41, 73
  etymology of, 25–26
  four elements of scientific discourse, 30–31
  knowledge defined as, 23–25, 85
  laboratory experiments, 40
  laws and rules of, 40–41, 73, 84
  military use of (*See* military use of science and technology)
  objectivity of, 39
  rationality of, 88
  relationships and, 26–28
  separation and, 25–26
  truth as goal of, 30–32, 40
Scott, James C., 210
security, 140–41, 156, 217. *See also* surveillance
  acceptance of authority for, 195
  citizen participation in, 159
  commercialization of, 185, 189–90
  as death-wish, 193
  fear and, 127, 142

security (*continued*)
  government information, secrecy of,
    176–77
  incentives to provide information, 141
  paranoia, effect of, 78
  privacy and, 29–30, 43–45, 174–76
  privatization of, 151, 175
  profits from, 160, 163, 167
  as slavery, 197
Seisint, 160–62
separation, 25–26
  fragmenting of the public, 43
  nature and technology, 209–10
Shagnon, Denis, 118
smart cards, 149
Smart Dust, 177–78
Smart Truck, 152–53
social pyramid, 190–93
Soldier Nanotechnologies, Institute for,
  148
South Africa, 220
Space Imaging company, 179
standardization, 79–80
status quo, 39
Stenbit, John, 187–88
students. *See* children
Sunshine Project, 41–42
surveillance, 8, 28, 113, 117, 126, 131. *See
  also* airport security; individuals, information on; monitoring of individuals;
  Panopticon
  biometrics, 117–18
  Carnivore packet analyzer used for,
    171–74
  of customers, 154–56
  definition, 132
  ECHELON computer software used for,
    169–70
  goods and services for, 148–50
  military use of, 182
  profits from, 160, 163
  unequal relationships as basis of, 26
  urban surveillance system, development
    of, 36
Sybase, 165
systems, 145

T

technocracy, 39, 110
technology, 85, 108, 110. *See also* biometrics; biotechnology; fingerprinting;
  nanotechnology; panoptic sort; RFID
  (Radio Frequency Identification) tags;
  robots
  acceptance by population, 141, 193–95,
    197–98
  bureaucracy and, 125, 132
  emotion-dead, 92
  etymology of, 82
  global positioning system, 122, 124
  government-subsidized technologies,
    five-step pattern in, 71–72
  as industrial production tool, 79
  laws and rules of, 40–41
  Marx and Engels and, 95
  military use of (*See* military use of
    science and technology)
  nature, separation from, 28, 209–10
  Panopticon use of, 181–86, 195
  power and, 39, 182
  regulation of, 174–76
  resistance to, 210–12, 214–15
  surveillance, 128
Teledata Communications company, 51
telephones, cellular, tracing location of,
  121–22
terrorism, 127, 150, 157–58. *See also* Total
  Information Awareness (TIA)
  air passenger screening, 128–31
  commercialization of, 160, 163, 185
  international cooperation to control,
    173–74
  power gained through, 90–91
Terrorism Information and Prevention
  System (TIPS), 201
Terrorism Information Awareness. *See*
  Total Information Awareness (TIA)
Tesco, 6, 153
thinking, standardized, 79–80
Thomas, Ned, 14–15
Thompson, Larry, 167
TIA. *See* Total Information Awareness
  (TIA)

TIPS (Terrorism Information and Prevention System), 201
tires, RFID chips in, 151–52
Torch Concepts, 129
Torpey, John, 147
torture, 140, 158
Total Information Awareness (TIA), 17–19, 148, 159–62, 188
TransUnion, 164
Traven, B., 112, 133–34
Tuskegee Syphilis Study, 46

U

unequal relationships, 26–27
UNICOR, 150
U. S. Air Force, 48–49, 178–79
U. S. Army, 148–49, 152–53, 182–83
U. S. Central Intelligence Agency (CIA), 121, 148, 167, 215. *See also* interrogation techniques
U. S. Combined DNA Index System, 121
U. S. Department of Commerce, 183
U. S. Department of Defense, 6, 29, 129, 187–89. *See also* Defense Advanced Research Projects Agency (DARPA)
   Afghanistan surveillance photographs, control of, 179
   interrogation (*See* interrogation techniques)
   National Nanotechnology Initiative, 69
   urban surveillance system, development of, 36
U. S. Department of Health and Human Services, 176
U. S. Department of Homeland Security, 149, 166
U. S. Department of Justice, 159, 166–67, 176. *See also* Ashcroft, John
U. S. Environmental Protection Agency, 71
U. S. immigration identification regulations, 117, 150

U. S. Information Awareness Office, 18–19, 23–24
U. S. Internal Revenue Service, 176, 215
U. S. National Institute of Justice, 42
U. S. Secret Service, 161–62
U. S. Senate Church Committee, 215
U. S. Transportation Security Agency (TSA), 128–30
United Nations, 118, 198–200
University of California, 177, 183
University of Southern California, 145–46
unmanned aerial vehicles (UAVs), 152–53, 178–82
urban surveillance system, development of, 36
USA Patriot Act. *See* Patriot Act

V

values, 96, 109
   of indigenous peoples, 82
   in panoptic sort, 112–13
Vetronix Corporation, 123
vivisection, 104–105

W

Wachowski brothers, 221–22
Wal-Mart, 6, 16, 97, 153
Washington, George, 11–12
Watson, David, 57, 205
wealth, 39, 142
   accumulation, 96–97
   bureaucracy and, 103
   concentration, 100, 140
Weber, Max, 92, 96–97, 100–104, 133, 214
welfare recipient identification, 119–20
Whitaker, Reg, 132
Winston Capital Fund, 166
Winston Partners, 165–66
working class, 113
World Summit on the Information Society, 198–200

About the Authors

George Draffan has been a carpenter, a corporate librarian, and a volunteer forest activist. He works as a freelance researcher and writer. He is the author or coauthor of *Railroads & Clearcuts, Strangely Like War,* and *The Elite Consensus,* and the forthcoming *Primer on Corporate Power.* Some of his work can be seen at www.endgame.org.

Derrick Jensen is the author of many books, including *A Language Older Than Words, The Culture of Make Believe,* and *Walking on Water. The Culture of Make Believe* was one of two finalists for the 2003 J. Anthony Lukas Book Prize. You can see more of his work at www.derrickjensen.org.

Chelsea Green is committed to being a sustainable business enterprise as well as a publisher of books on the politics and practice of sustainability. This means reducing natural resource and energy use to the maximum extent possible. We print our books and catalogs on chlorine-free recycled paper, using soy-based inks, whenever possible. *Welcome to the Machine: Science, Surveillance, and the Culture of Control* was printed on Legacy Trade Book Natural, a 100 percent post-consumer waste recycled, old growth forest-free paper supplied by Webcom.

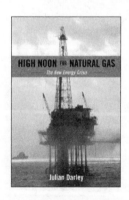